D1536644

Made In Greece

A Guided Tour of Greek Cooking

John S. Kaldes

PUBLISHINGWORKS
2004

Printed in Canada.

First printing.

Cover and interior design by Greg Zompolis.

Published by:

PublishingWorks
4 Franklin Street
Exeter, New Hampshire 03833
800/333-9883
www.publishingworks.com

www.MadeinGreececookbook.com

Distributed to the trade by Enfield Distribution

Library of Congress Cataloging-in-Publication Data
Kaldes, John S.
    Made in Greece : a guided tour of Greek cooking / John S. Kaldes.
        p. cm.
    Includes index.
    ISBN 0-9744803-6-3 (alk. paper)
    1. Cookery, Greek. 2. Menus. I. Title.
TX723.5.G8K35 2004
541.594945--dc22   2004053164

Table of Contents

Acknowledgments

This book is dedicated to:

My wife Zoi, who supported me in the creation of this book, and who never complained about the mess I was leaving behind in her kitchen.

To my son Peter, who helped me with the initial edit and acted as my official "taste-tester" by trying all of my recipes.

To my son Gregory, who made the Web page a reality.

And to all of my friends, who encouraged me to write this book, and to those wonderful Greek cooks who shared some of their secret recipes with me.

PREFACE

This book is for those who have experienced authentic Greek cuisine through their travels, or at a local Greek restaurant, a Greek festival, or a friend's house. It's also perfect for the adventurous diner who enjoys trying something new and different, and for those who may have seen a Greek recipe on television and would like to experiment with a healthier way of preparing authentic Greek foods. There are some good Greek cookbooks available today, but most do not address the specific problems facing the vast majority of people who are not professional cooks—such as availability of ingredients, ways to maintain authentic cuisine flavor without the calories and fat, and complicated preparation. Unfamiliar recipes are often made with hard-to-find ingredients and many don't clearly say how to serve the dish or what the finished recipe should look like.

Made In Greece eliminates the intimidation of cooking unfamiliar recipes by using easy-to-find ingredients, explaining which tool to use, and providing step-by-step instructions for every recipe. This is a cookbook that can be used by both professional and novice cooks who want to learn or experiment with new tastes, or by those who want to impress their friends by cooking a special Baklava, a fabulous spinach pie, or an incredibly tender lamb Souvlaki.

The recipes here are simple and the results are splendid. Ambiguities have been removed by providing clear directions, sequential instructions, and cooking times. The vast majority of the ingredients are available at the local supermarket. There are a couple of ingredients, like mahlepi or mahleb and mastich chiou, which can be easily found in Greek food stores or can be ordered from the Internet. To make it easier for those who have no easy access to some of the ingredients, there are recommended substitutes.

Cooking is an art. Try the recipe, and if you wish, experiment with different ingredients. Substitute what you like or what you think you will like. This book is your guide to exploring and enjoying the beautiful and excellent foods of the island of Mytilini, Greece.

INTRODUCTION

This is the first book I have ever written. I must admit it seemed quite daunting when I first set out to write this cookbook, especially since so many other cookbooks are available. Throughout my research, I discovered few Greek cookbooks that provide truly authentic Greek recipes. Many of my friends have asked me how and where they could get recipes for the dishes that I have prepared for them at home. They have asked me how to make Baklava or Spanakopita, without the filo dough falling apart or sticking together, and for many other cooking tips. It is always a pleasure for me to share my cooking secrets with friends and with all who love authentic Greek food. So, it is with great pleasure that I offer you this cookbook of delicious, authentic Greek recipes.

I was born in Plagia, a small village on the Aegean island of Mytilini, Greece. I didn't cook much until I met my father-in-law, who inspired me. I have learned so much from him over the past twenty-five years. I also spent many hours watching my mother and sister prepare tantalizing Greek dishes, and it is from them that I learned many of the traditional techniques and recipes offered in this book. Today, cooking is my passion.

I have written this book in an easy-to-follow format to help anyone—even novice cooks—to learn to cook superb, authentic Greek dishes. The majority of these recipes are simple. Even the seemingly complicated dishes, such as Dolmadakia, Mousaka, and Spanakopita, become much easier using my instructions.

Today, more than ever, people are conscious of the importance of eating healthy foods. There is a new demand for low-fat foods and for nutritious and tasty low-fat recipes. I have tried to create all of the recipes in this book using low-fat ingredients, including substituting olive oil for butter. There are, however, instances where substituting butter does not do justice to the recipe, so in a few of my recipes, I use butter to preserve the taste. It is up to each one of us to eat everything in moderation.

What I offer to you are great recipes, healthy ingredients that you can find anywhere, simple, clear instructions, fantastic results every time, and a way to make Greek cooking fun.

Why Cook with Olive Oil?

Growing up, I did not know anything about butter. My father was an olive grove farmer, so butter was not allowed at home. My mother would only use butter when she made desserts, but that was the exception. She cooked everything using the freshest and best olive oil around.

Using fresh, high-quality olive oil applies to all traditional Greek cuisine. You can't think Greek cuisine without olive oil. Olive oil and Greek food are synonymous. Even one drop of this golden liquid provides the harmony of flavors and the simplicity you find in authentic Greek foods.

Cooking with olive oil is also very healthy. Many studies published in the last few years indicate that it is a lot healthier to use olive oil for cooking than it is to cook with butter or other fats. In addition to healthy eating, high-quality olive oil will enhance the flavor of your food and add unparalleled aromas to the various dishes.

Greeks use fresh, raw, olive oil in their salads, soups, sauces, dips, and even to wet their crusty bread. It is the main ingredient for vegetarian "ladera" dishes made without meat or fish. It is also used in meat, chicken, and fish dishes as well. In general, Greek cooks use it to cook most of their foods.

In Greece, each Greek consumes about forty pounds of olive oil per year, compared to about one pound a year for the U.S. population.

The island of Mytilini cultivates about 15 million olive trees and produces about 30 million pounds of olive oil every year. This is about 15 percent of the total olive oil production of Greece. This, of course, varies from year-to-year. Olive trees live a very long time, so it is not unusual to find trees in Greece that are 300 years old or older.

Oil Classifications

Olive oil is produced from the fruit of the olive tree. It is considered pure if it is produced using a mechanical process that does not alter the oil in any way. No chemicals are introduced in the process. The only treatment olives receive after they are picked from the tree is storage, washing, crushing, centrifuging, and filtering.

Olive oils have different designations and classifications depending on their acidity, taste, aroma, and color. Acidity (free fatty acids) is the main determinant in classifying and pricing the oil, with taste being the second determinant.

Extra Virgin Olive Oil

This is the best of the olive oils with acidity not to exceed 1 percent. The low acidity, combined with an excellent fruity taste, aroma, and a beautiful color make it absolutely perfect for salads, soups, dips, and sauces.

Virgin Olive Oil

Acidity for this oil will not exceed 2 percent. This oil has a fine taste, aroma, and color. It is used for baked dishes, vegetable dishes, for frying fish, potatoes, omelets, and for bread making. There is also a semi-fine virgin olive oil with an acidity of less than 3.3 percent. This oil has a good flavor and aroma. It is used most often for frying.

Virgin Olive Oil (Lampante)

The acidity of this oil is more than 3.3 percent with noticeable defects in taste, aroma, and color. This oil will normally be made using a chemical process and other refinements.

Olive Oil

This is a blend of refined virgin olive oil (lampante) and virgin olive oil (fine or semi-fine). This oil normally has an acidity of less than 1.5 percent. This is the most common and least expensive olive oil available in the market.

There are other terms of oil classification, but none is officially regulated or recognized. In my opinion, the new methods of oil processing have rendered the terms "cold- pressed" or "first-press" olive oil obsolete, especially for mass-produced olive oil.

Factors dramatically affecting the quality of olive oil are: ripeness of the fruit, the harvesting method used, disease, storage, time between harvesting and processing, temperature, and other factors which are less significant.

Oxygen, light, and heat also have significant impact on the oil after its production. Protect your olive oil from extreme light and heat to prevent oxidation, which will render olive oil unusable.

SUGGESTED MENUS

Below you will find some suggested menus for various occasions. I hope you will find these helpful in preparing your authentic Greek meal. Do not hesitate to develop your own menus using the recipes that are your favorites. After all, we should always cook and eat what we like. These are some of the menus I prepare when I cook at home.

Christmas Day

Feta Cheese Pie, Potato Croquettes, Russian Salad, Boiled Shrimp
with Lemon Sauce
Various Greek Cheeses

Stuffed Turkey
Oven Roasted Potatoes, Green Salad

Butter Cookies Covered with Confectionary Sugar, Honey-Dipped Cookies
Baklava with Walnuts
Coffee

Easter Dinner

Baby Beef Liver with Oregano, Spinach Pie, Eggplant Dip, Eggplant
with Yogurt Sauce
Various Greek Olives and Cheeses
Pita Bread

Stuffed Shoulder of Lamb, Souvlaki (Lamb)
Rice Pilaf, Oven Roasted Potatoes, Mixed Green Salad

Easter Bread, Red Colored Eggs
Baklava with Almonds, Butter Cookies with Mastich
Coffee

Winter Party Menu

Bean Soup
Various Greek Olives, Scallions, Red Radishes
Crusty Bread
Baked Salted Codfish, Cabbage Salad,
Farina Cake with Syrup, Coffee

Spring Party Menu

Meat Balls, Cheese Puffs, Fried Artichokes
Chicken with Egg-Lemon Soup
Lamb with Oregano
Dandelion Salad
Walnut Cake, Coffee

Summer Party Menu

Fried Sand Shark, Garlic Dip, Zucchini Croquettes, Cheese Saganaki
Stuffed Squid with Rice, Grilled Shrimp
Stuffed Tomatoes with Rice
Greek Salad
Shredded-Dough Walnut Pastry, Coffee

Fall Party Menu

Pumpkin Fritters, Stuffed Grapevine Leaves, Caviar Croquettes
Greek Cheeses
Mousaka
Custard Pastry, Coffee

Family Menu 1

Stuffed Mushrooms, Taramasalata
Pita Bread
Chicken with Tomato and Olives
Rice Pilaf
Yogurt, Honey, and Walnuts
Coffee

Family Menu 2

Beans with Tomato Sauce
Bread
Stuffed Chicken
Green Salad
Babas, Coffee

Family Menu 3

Chickpea Soup
Country Style Bread
Pork with Lettuce
Farina Cream Cake, Coffee

Family Menu 4

Lentil Soup
Pita Bread
Fried Flounder Filet
Cake with Apricot Preserves
Coffee

Handy Measurements and Conversion Tables

Oven Temperatures	Volume Measurements
275° F = 140° C	1/2 fl oz (fluid ounce) = 55 ml (milliter)
300° F = 150° C	3 fl oz = 75 ml
325° F = 170° C	5 fl oz (quarter pint) = 150 ml
350° F = 180° C	10 fl oz (half pint) = 275 ml
375° F = 190° C	3/4 pint = 425 ml
400° F = 200° C	1 pint = 570 ml
425° F = 220° C	1 1/4 pint = 725 ml
450° F = 230° C	1 1/2 pint = 1000 ml (1 liter)
2 pint = 1.2 liter	

Weight Measurements

1/4-ounce = 7 g (grams)	5 ounces = 150 g
1/2-ounce = 10 g	6 ounces = 175 g
3/4-ounce = 20 g	8 ounces = 225 g
1 ounce = 25 g	9 ounces = 250 g
1 1/2 ounces = 40 g	10 ounces = 275 g
2 ounces = 50 g	12 ounces = 350 g
2 1/2 ounces = 60 g	16 ounces (1 pound) = 450 g
3 ounces = 75 g	2 pounds = 900 g
4 ounces = 110 g	2.2 pounds = 1000 g (1 kilo)

Other handy measurements for various items used in this book:

1 teaspoon liquid = 5 ml

1 tablespoon liquid = 15 ml

4 tablespoons liquid = 1/4 cup

1 cup liquid = 200 ml

1 cup flour = 6 ounces

1 cup rice = 8 ounces

1 cup sugar = 8 ounces

1 lemon, juice of = 4 tablespoons

3 teaspoons = 1 tablespoon

1 egg (large) = 2 ounces

Length Measurements

1/8 inch = 3 mm (millimeters)

1/4 inch = 5 mm

1/2 inch = 1 cm (centimeter)

3/4 inch = 2 cm

1 inch = 2-1/2 cm

1 1/4 inches = 3 cm

1 1/2 inches = 4 cm

1 3/4 inches = 4-1/2 cm

2 inches = 5 cm

3 inches = 7-1/2 cm

5 inches = 13 cm

6 inches = 15 cm

8 inches = 20 cm

9 inches = 23 cm

10 inches = 25-1/2 cm

11 inches = 28 cm

12 inches = 30 cm

Meat Roasting Chart

Meat

Internal temperature when removed from
 oven (in degrees Fahrenheit)

Beef

Rare: 120–125

Medium rare: 130–140

Medium: 145–150

Well done: 155–165

Lamb

Rare: 130–135

Medium rare: 140–145

Medium: 150–160

Well done: 160–165

Pork

Fresh: 140–150

Poultry

Chicken: 180–185

Turkey: 180–185

NOTE: The internal temperature will rise by about 2–3° F after the roast is removed from the oven.

Equipment

I have attempted to keep the equipment needed to prepare my recipes to a minimum. You do not need a large number of tools, but rather, the correct tools. It does not matter for whom you are cooking the meal, but it is essential that you use the proper equipment so that the results will be rewarding. I do not think that you should go out immediately and buy the best kitchenware, but make it a habit to purchase a few items at a time, and you will eventually be able to acquire a nice set of cooking utensils that will meet your cooking needs.

Basic Tools

Forks, spoons, knives, spatulas, tongs, grater, sieves, measuring cups, scales, timer, thermometer, needle, cotton string and skewers, food mill, food processor, electric mixer, rolling pin, pastry brushes, pastry cutters, kitchen towels, cutting boards.

Different sizes of high quality, sharp knives are essential. No matter what level of cooking you do, you should always have a good set of knives.

Wooden kitchen forks are ideal for fluffing rice, and wooden spoons and spatulas are ideal for mixing in nonstick saucepans. This way you will not damage your pans.

Also, a collection of metal spoons and forks, including a large, perforated ladle, and solid and flexible spatulas, are essential. Brushes are also essential, especially when using filo pastry to make spinach pie, cheese pie, baklava, etc.

Good quality kitchen tongs are another essential when grilling steaks, vegetables, or even fish. You need tongs with teeth and also a set of flat tongs. One four-sided grater will do for all your needs.

If you prefer to grind your own meat, then I suggest that you purchase a manual or electric meat grinder. Make sure the grinder you choose has different sized cutters.

Try to purchase metal sieves of two to three different sizes. They are not expensive, but come in very handy when you want to strain broth, sieve flour, etc. A colander is another item you will use frequently to drain various vegetables or spaghetti.

A measuring cup is an absolute essential. Purchase a glass measuring cup with a 2- to 4-cup capacity with volumes marked on the outside of the glass.

When you bake desserts, it is essential to use a scale; one that measures up to 5 pounds will do well.

A thermometer and timer are also essential. Sometimes you will need to check your meat temperature, so I recommend that you keep a meat and oven thermometer handy. A timer with a loud, audible alarm is also nice to have. Get one that attaches to your refrigerator door, so that it is always handy and easy to reach and program.

Cotton string, skewers, and a needle are also a few of the things you may need when you plan to stuff a turkey, chicken, lamb, or to grill Souvlaki or shrimp.

A food mill is also a very good and inexpensive tool to have. It is invaluable when you make vegetable soup, mash potatoes, etc. A food processor is expensive, but is essential if you will be shredding, slicing, and grating frequently. It is also helpful in preparing Taramasalata and Skordalia, or to grind walnuts to make walnut cake.

An electric mixer is also an essential tool to have.

When you plan to make cookies or open your own filo pastry, a rolling pin and one or two sets of different style pastry cutters are essential.

Also, a few kitchen towels will be needed, especially when you make bread.

I also recommend a couple of good cutting boards: one small board for little everyday jobs and a larger one for the bigger jobs. Make sure you clean them well with soap and water after every use.

Saucepans
Saucepans are the most frequently used equipment in this book. There is a huge variety of saucepans in the market. I prefer and recommend using stainless steel saucepans with good, heavy, heat-conductive bases. These saucepans are easy to keep clean and easy to handle. I recommend having two small saucepans: one pint and one quart; two to three medium saucepans (2 quarts); and one or two large saucepans (6 to 8 quarts).

Frying Pans
I prefer the nonstick metal ones, but you will need to be very careful not to use metal utensils. If you do, you will need to replace them very quickly, as they will become very difficult to clean. You will need one frying pan with an 8-inch base and one with a 10-12-inch base.

Baking Pans and Roasting Pans

Do not purchase the thin-gauge, aluminum ones. Use heavy-duty, stainless steel pans. The most common sizes of baking pans used in this book are a 12 x 9 x 2-inch or 15 x 10 x 2-inch pan, but having several pans of different sizes is a very good idea. Consider also purchasing a 9-inch pie pan, a 10 x 1-inch round pan, and a 12 x 1-inch round pan, as you will find that these will also come in handy.

Baking Sheets

You will need a couple of baking sheets, especially when you make cookies or crispy Baklava. I prefer a 12 x 17 x 1-inch baking sheet.

Mixing Bowls

A set of small, medium, and large, plastic mixing bowls are essential for every kitchen. It is also nice to have medium-sized metal, mixing bowls to use with your electric hand-mixer to beat eggs or other ingredients.

APPETIZERS

In Greece, appetizers, or meze, sometimes become the main meal. Greeks like to talk about politics or sports while enjoying their ouzo or wine. They consume so many appetizers that oftentimes, the main course becomes redundant. Typically, when you go to a restaurant in Greece, you will get the menu, but it does not do justice to what they have to offer. That is why you will see the waiter come out with large trays of appetizers and explain to you what they are and how they are made. Now you have the choice of selecting a few meze with which to start your meal. However, do not overindulge yourself, as there will be more food coming. Keep in mind that after the appetizers, a substantial main meal and dessert will follow.

In this chapter, you will find a variety of appetizers, which represent a small sample of what is available. The appetizers in this book are those that are the most familiar to people outside Greece, so when you visit Greece, and especially the island of Mytilini, you will be able to identify all of them in the various restaurants or tavernas.

Spinach Pie.

Spinach Pie (Spanakopita)

This is classic Greek pie. It is one of the best-known Greek dishes throughout the world. It is so popular that it is served in every Greek restaurant and every church bazaar.

Pastry

> 1 pound filo pastry #4
> 1/2-cup extra virgin olive oil

Filling

> 20 ounces fresh spinach, chopped or 20 ounces frozen, chopped baby leaf
>   spinach
> 1/4-cup extra virgin olive oil
> 1 large onion (8 ounces), finely chopped
> 1 cup green scallions, finely chopped
> 1 pound Feta cheese, grated
> 4 eggs, beaten
> 3 tablespoons fresh parsley, finely chopped
> Pepper

Remove filo pastry from refrigerator. Leave at room temperature for 6 hours. If frozen, thaw at room temperature for 12 hours. Keep in original packaging. Drain thawed spinach before using.

To Prepare Filling

Heat oil in a medium saucepan. Add onions and sauté until soft. Add scallions; sauté for 1–2 minutes. Add spinach to onion mixture; sauté until juices evaporate. Remove from heat and let cool for 5 minutes.

Add Feta cheese, eggs, parsley, and pepper to spinach mixture. Mix to blend ingredients. Mixture should be soft without any liquid. Set aside.

Remove 20 filo pastry sheets from packaging and unroll them onto a kitchen towel. Measure filo sheets and cut to fit flat into a 12 x 9 x 2-inch baking pan. Keep filo sheets covered with towel to prevent drying. Grease baking pan with olive oil. You will use 10 filo sheets under and 10 on top of the spinach mixture.

Lay 1 filo pastry sheet flat in greased pan and brush top generously with oil; add a second filo sheet and brush with oil. Repeat process with remaining 8 filo sheets. Pour spinach mixture into baking pan and spread out evenly. Place a filo sheet on top of the mixture, brush with oil, and continue layering and brushing with oil until all remaining filo sheets are used. Brush last sheet with oil; ensure edges are well oiled.

Refrigerate pie for 15 minutes. Preheat oven to 350° F. Remove pie from refrigerator. With a sharp knife, cut through the filo sheets into 3 x 3-inch square pieces. Place pan on middle oven rack; bake for 45–50 minutes or until golden. Turn oven off and let the pie sit in the oven for an additional 10 minutes.

Remove pie from the oven and set aside for 10 minutes to cool. Serve warm or cold.

Tip: You will use three-quarters of the filo pastry for this recipe (there are approximately 25–27 sheets per package). You can use trimmings between the layers: cut them into small pieces, spread them over the existing sheets and brush with oil. Repeat layering until all are used. You can also use all of the sheets; just place half on the bottom and half on the top.

Preparation time: 45 min.   Cooking time: 1 hr.   Yield: 12 pieces

Feta Cheese Pie
(Tyropita)

This is very popular dish. You can serve it hot or cold, as an appetizer or as a main course.

Pastry
>   1 pound filo pastry # 4
>   1/2-cup extra virgin olive oil or 8 ounces butter, melted

Filling
>   1 pound Feta cheese, grated
>   1/4-cup milk
>   3 tablespoons parsley, finely chopped
>   1 tablespoon dill, snipped
>   3 eggs, beaten
>   Pepper

Remove filo pastry from refrigerator and keep at room temperature for 6 hours. If frozen, thaw at room temperature for 12 hours. Keep in original packaging.

To Prepare Filling
In a medium mixing bowl, mix Feta, milk, parsley, dill, eggs and pepper; set aside. Remove 20 filo pastry sheets from original packaging and unroll on a kitchen towel. Measure filo sheets and cut to fit a 12 x 9 x 2-inch baking pan. Keep filo sheets covered with kitchen towel to prevent from drying. Using oil, grease baking pan.

Place 1 filo sheet in pan and brush entirely with oil; add second filo sheet and brush with oil. Repeat process with remaining 8 filo sheets. Pour cheese mixture into pan and spread out evenly. Place another filo sheet on top of the mixture, brush with oil, and continue layering and brushing with oil until all remaining filo sheets are done. Brush last sheet with oil, ensuring edges are well oiled.

Refrigerate 15 minutes. In meantime, preheat oven to 350° F.

Remove from refrigerator. With a sharp knife, cut through the filo sheets to make 3 x 3-inch square pieces.

Bake on the middle rack for 45–50 minutes or until golden. Turn oven off and let pie sit in the oven for another 10 minutes.

Remove from oven and set aside to cool for 5 minutes. Serve warm or cold.

Tip: Freezes well when uncooked. If frozen, bake for 1 hour or until golden. You will be using about three-quarters of the filo sheets to make this recipe (approx. 25–27 sheets per package). You can use trimmings between the layers by cutting them into small pieces, spreading them over the existing sheets and brushing them with oil. Repeat layering until all are used. You can also use all of the filo sheets; just place half on the bottom and half on the top.

Preparation time: 35 min.   Cooking time: 1 hr.   Yield: 12 pieces

Leek Pie
(*Prassopita*)

Not as well known as spinach pie, but popular in wintertime when leeks are abundant.

Pastry
>1 pound filo pastry #4
>1/2-cup extra virgin olive oil

Filling
>2-2 1/2-pounds leeks
>2 medium onions, finely chopped
>2 tablespoons extra virgin olive oil
>1/4-cup parsley, coarsely chopped
>4 eggs, beaten
>1/4-cup milk
>4 ounces Feta cheese, grated
>Pepper

Remove filo pastry from refrigerator. Keep at room temperature at least 6 hours. If frozen, thaw at room temperature for 12 hours. Keep in original packaging.

Filling
Discard outside layer from the leeks, slice leeks lengthwise and cut into 1/2-inch wide rounds. Wash 3 to 4 times in cold water.

Bring 5 cups of water to a boil in medium saucepan. Add onions and leeks; blanch for 1 minute. Drain well. Set aside. Wipe saucepan clean. Sauté leeks and onions in 2 tablespoons of oil for 2 minutes. Reduce heat to very low, cover; steam for 10 minutes or until soft, mixing occasionally. Remove from heat; add parsley, mix, and set aside for 5 minutes to cool.

Add eggs, milk, cheese, pepper, and mix lightly to incorporate all the ingredients. Remove 20 filo sheets from package; unroll onto a kitchen towel. Measure filo sheets and cut to fit a 12 x 9 x 2-inch-baking pan. Keep sheets covered with towel to prevent drying. Grease baking pan with olive oil.

Place 1 filo sheet in pan, brush generously with oil. Add second sheet; brush with oil. Repeat process with remaining 8 sheets. Pour cheese mixture into pan; spread out evenly. Place a filo sheet on top of the mixture, brush with oil, and continue layering and brushing with oil until remaining 9 sheets are used. Brush last sheet with oil, ensuring edges are well oiled. Refrigerate for 15 minutes. Preheat oven to 350° F.

Remove from the refrigerator. With sharp knife, cut sheets into 3 x 3-inch squares. Place on middle oven rack; bake for 45–50 minutes or until golden. Turn oven off and let pie sit in the oven for an additional 10 minutes.

Remove from oven and set aside for 10 minutes to cool. Serve warm or cold.

Tip: Select leeks with long white stems; use only tender portion of the green plant. Pie freezes well uncooked. If frozen, thaw and bake for 60 minutes or until golden. You will use about three-quarters of the filo pastry to make this recipe (approx. 25-27 sheets per package). To use trimmings, cut into small pieces, spread over existing sheets, and brush with oil. To use all of the sheets, place half on the bottom and half on the top.

Preparation time: 45 min.   Cooking time: 1 hr.   Yield: 12 pieces

Chicken Pie
(*Kotopita*)

This is another way to surprise your friends with a delicious meal. It is a creamy dish that makes an ideal quick lunch.

Pastry

> 1 pound filo pastry #4 or #7
> 1/2-cup virgin olive oil plus 3 tablespoons

Broth

> 2 pounds chicken breasts
> 1 medium onion, whole
> 1 celery stalk, cut in half
> 1 carrot, cut in half
> 2 bay leaves
> 10 peppercorns

Filling

> 2 tablespoons flour
> 1 cup chicken stock, hot
> 1 cup light cream, warm
> 2 eggs, beaten
> 1 teaspoon ground allspice
> 6 ounces Feta cheese, crumbled
> 1/2-cup dill, finely chopped
> 2 tablespoons fresh sage, chopped finely
> Salt, pepper

Remove filo pastry from refrigerator. Keep at room temperature for 6 hours. If frozen, thaw at room temperature for 12 hours. Keep in original packaging.

To Prepare Chicken and Broth

Wash chicken thoroughly. Place chicken breasts in large saucepan; cover with water (about 6 cups); bring to a boil. Add onion, celery, carrot, bay leaves, and peppercorns; cover and simmer until chicken is soft (about 45–50 min.). Remove from heat; drain. Preserve chicken and 1 cup stock. Discard vegetables, skin, and bones. Chop chicken breasts into small cubes and set aside.

To Make Filling

In medium saucepan, heat 3 tablespoons oil; add flour, mixing frequently until light golden. Slowly add hot stock; stir vigorously to dissolve lumps. Reduce heat, add cream, and stir. Remove sauce from heat and set aside for 10 minutes to cool.

Add chicken cubes, eggs, allspice, cheese, dill, sage, salt, pepper; stir to coat chicken with the sauce and blend all the ingredients.

Remove 20 filo sheets from package; unroll onto a kitchen towel. Measure sheets and cut to fit 12 x 9 x 2-inch baking pan. Keep covered with towel to prevent drying. Grease baking pan with olive oil.

Place 1 filo sheet in baking pan; brush generously with oil; add second sheet; brush with oil.

Repeat process with remaining 8 sheets. Pour chicken mixture into pan; spread out evenly.

Place another filo sheet, brush with oil. Continue layering remaining sheets. Brush last sheet with oil; ensure all edges are well oiled. Refrigerate for 15 minutes. Preheat oven to 350° F.

Remove pie from refrigerator. With sharp knife, cut filo sheets into 2 x 3-inch or 3 x 3-inch squares. Bake on the middle oven rack for 45–50 minutes or until golden. Turn oven off and let pie sit in the oven for an additional 10 minutes.

Remove from oven and serve warm. Serve as a starter or main meal.

Preparation time: 1 hr. 10 min.   Cooking time: 1 hr.   Yield: 12–15 servings

Meatballs
(Keftedes)

This is a very versatile dish that can be served as an appetizer or main course. Delicious hot or cold.

  1 pound ground beef
  1/2-cup milk
  4 ounces dry bread crumbs
  1 medium onion, grated
  1 egg
  1/4-cup white wine
  3 tablespoons parsley, finely chopped
  3 tablespoons ouzo or brandy (optional)
  2 teaspoons ground cumin
  1 tablespoon dry oregano
  Salt, pepper

For Frying
  1 cup virgin olive oil
  1 cup flour

In medium mixing bowl, pour milk and soak bread crumbs. Add ground beef, onion, egg, wine, parsley, ouzo, cumin, oregano, salt, and pepper; mix well.

Make meatballs the size of a large walnut. Shape into small patties and toss in flour.

In a frying pan, heat half of the olive oil. Fry meatballs on both sides until golden. Add more oil as required.
Serve hot or cold.

Tip: Substitute ground lamb or mixture of ground beef and lamb for the ground beef. To keep meatballs from sticking to your hands, periodically wet your hands with water. To serve them as an appetizer, make each meatball the size of a large cherry.

Preparation time: 20 min.   Cooking time: 25 min.   Yield: 28–30 meatballs.

Feta Cheese Dip
(Kopanisti)

This is a nice dip for any occasion. It combines Feta cheese, cream, and a sprinkling of extra virgin olive oil.

    1/4-pound Feta cheese, soaked in milk for 12 hours
    3 tablespoons light cream
    1 tablespoon fresh parsley, chopped
    Pepper
    1 tablespoon extra virgin olive oil

Put cheese and light cream in a food processor. Blend for 4-5 seconds until smooth. Add parsley and pulse 3–4 times. Add pepper and pulse 3-4 times. Remove from blender and place into a serving bowl; refrigerate.

When ready to serve, pour oil over the cheese and serve with pita bread wedges.

Preparation time: 10 min.   Yield: 1 cup

Potato Croquettes
(Patatokeftedes)

This recipe combines boiled potatoes with cheese and eggs cooked in olive oil. It is excellent as an appetizer or as side dish for a variety of meat dishes.

> 1 pound potatoes
> 4 ounces Kasseri or Kefalotyri cheese, shredded
> 4 ounces Feta cheese, grated
> 2 egg yolks
> 3 tablespoons milk, cold
> 1 teaspoon nutmeg
> Salt, pepper
> 3-4 tablespoons instant mashed potatoes

For Frying
> 1 cup virgin olive oil
> 3 tablespoons flour

Boil potatoes with skin. When soft, remove from heat, drain, and set aside to cool. Peel and puree potatoes in a food mill.

Place pureed potatoes into a medium-mixing bowl. Add cheeses, egg yolks, milk, nutmeg, salt, pepper, and mix well. Slowly add mashed potatoes until mixture is firm; you want a very firm mixture.

Using a tablespoon, scoop up enough of the potato mixture to make croquet the size of a large walnut. Toss each croquette in flour and shape into small sausages about the size of your thumb. See Tip below before tossing the croquettes in the flour.

In a medium frying pan, heat half of the oil. Oil should not be very hot. Fry potato croquettes until golden (1–2 minutes on each side).

Remove from heat and place on paper towel to remove any excess oil.

Serve hot or cold.

Tip: You may substitute white cheddar cheese in place of Kasseri or Kefalotyri cheese. Consistency of the mixture is very important. When the croquette mixture is prepared, and before you toss individual croquettes in the flour, prepare and fry only one croquette to ensure that the mixture is the right consistency. If croquette splits, add more instant mashed potatoes to make the mixture thicker. You may have to try this a couple of times until you achieve the perfect consistency.

Preparation time: 50 min.   Cooking time: 20 min.   Yield: 30–35 croquettes

Creamy Caviar Dip
(*Taramasalata*)

Tarama is carp fish caviar and it is available in small jars at Greek and Italian groceries. This is a very popular dip in Greek cooking. It is served as an appetizer and goes well with ouzo. You must try this one. You will love it.

> 2 tablespoons Tarama
> 10 ounces (10 slices) plain, white Italian bread
> 1 cup extra virgin olive oil
> 1 lemon, juice of
> 1 very small onion
> 1 tablespoon fresh dill, snipped
> 5 black olives, pitted and sliced in half
> Pita bread cut into wedges

Remove all crust from bread and discard. Toast bread lightly just to dry up. Soak in water for a few seconds, then squeeze and drain all of the water out until the bread is like a ball of dough. Set aside.

In food processor, pulse Tarama 3–4 times.

Add wet bread into blended Tarama and pulse 4–5 times or until mixture is smooth. Gradually pour oil into mixture and continue blending until entire amount of oil is incorporated and the mixture is a smooth paste.

Grate onion and collect 2 teaspoons of the onion juice. Discard onion pulp. Add onion and lemon juice to Tarama mixture and pulse 2–3 times.

Remove and place into a serving dish. Cover with plastic wrap or foil and refrigerate for a few hours.

Garnish with olives and dill and serve with pita bread wedges.

Tip: Substitute crackers for pita bread. Cover and refrigerate leftover Taramasalata for one week.

Preparation time: 20 min.   Yield: 3 cups

Zucchini Croquettes
(Kolokythokeftedes)

This dish is very popular when fresh zucchini and mint are readily available. Mint gives a refreshing flavor to this dish.

2 pounds green zucchini
1 large onion, grated
1/2 red bell pepper, finely chopped
2 tablespoons virgin olive oil
1/2-cup Kefalotyri or Kaseri cheese, grated
2 eggs
1/4-cup fresh parsley, finely chopped.
3 tablespoons fresh mint, finely chopped
Salt, pepper
1 cup dry bread crumbs

For Frying
1/2-cup virgin olive oil
1/2-cup flour

Wash zucchini. In a medium mixing bowl grate zucchini and squeeze out water. Discard water.

In a small saucepan, add 2 cups water and bring to a boil. Blanch onions and bell pepper for 1 minute. Drain and discard water. Wipe saucepan dry. Heat 2 tablespoons of oil and sauté onions and red pepper until soft. Remove from heat, drain, and set aside to cool.

In bowl with zucchini, add onions, red pepper, and mix. Remove any excess water or oil. Add cheese, eggs, parsley, mint, salt, pepper, bread crumbs and mix well. If mixture is too watery, add more bread crumbs. Mixture must be firm.

Using a tablespoon to pick up the zucchini mixture, make balls the size of a small egg. Toss balls in flour and shape them into small patties. See Tip below before tossing in flour.

In a frying pan, heat half of the olive oil. Fry zucchini croquets over a medium heat for 2-3 minutes on each side or until golden. Remove from heat and place on paper towel to remove any excess oil.

Serve hot or cold.

Zucchini Croquettes (con't)

Tip: You may substitute Parmesan cheese for the Kaseri or Kefalotyri cheese. Ensure that all the ingredients are dry before you mix them together. It is important that the mixture's consistency be thick so that the patties will maintain their shape when put into the frying pan.

When the croquette mixture is prepared, and before you toss individual patties in the flour, prepare and fry only one to ensure that the mixture is of the right consistency. If the patty splits, then add one or two more tablespoons of bread crumbs to make the mixture thicker. You may have to try this a couple of times to achieve the perfect consistency.

Preparation time: 40 min.   Cooking time: 20 min.   Yield: 30–35 patties

Stuffed Mushrooms with Greek Sausage (*Manitaria Gemista*)

The Greek sausage, with its subtle garlic flavor, makes this dish very interesting. An excellent appetizer.

> 20 white mushrooms, medium 2 inches wide
> 12 ounces fresh Greek sausage
> 2 tablespoons extra virgin olive oil
> 1 small onion, grated
> 1/4-cup breadcrumbs
> 1 tablespoon dry oregano
> 1/4-cup white wine
> Salt, pepper
> 1 egg, beaten in 2 tablespoons milk
> 3 tablespoons fresh oregano, finely chopped

Rinse mushrooms with cold water and immediately pat dry with paper towel. Remove stems and set aside. With a small sharp knife, remove any remaining stems and enlarge the mushroom cap. Chop mushroom stems into small pieces and set aside.

Remove skin casing from sausage and break meat into small pieces. In a large skillet, heat oil and sauté onion until soft and golden. Add mushroom stems and sauté.

Add sausage meat and brown for 4–5 minutes, mixing frequently. Add bread crumbs, dry oregano, wine, salt, pepper, and mix. Simmer for 2 minutes or until juices are absorbed. Remove from heat and let it cool for 5 minutes. Preheat oven to 375° F.

Pour egg into sausage mixture and mix thoroughly to blend with stuffing. With a spoon, fill mushroom caps with stuffing. Use excess stuffing to make small balls and bake with mushrooms.

Arrange mushrooms in a baking pan, cover with aluminum foil, and bake on the middle rack for 20 minutes. Remove from oven and baste stuffing with juices. Sprinkle with fresh oregano and serve hot.

Tip: If Greek sausage is not available, you can substitute sweet Italian sausage, and add 1 small garlic clove, finely chopped, to skillet, and add 1/2 teaspoon ground anise seeds and 1 teaspoon of orange zest to sausage mixture.

Preparation time: 30 min.   Cooking time: 20 min.   Yield: 20 pieces

Fried Mushrooms
(Manitaria Tiganita)

This is a fantastic traditional Greek dish that is made using wild mushrooms that are found growing under the pine needles in the forests on the island of Mytilini. The mushrooms are picked in late November and the harvest lasts about three weeks. Farm-grown, white mushrooms are a very good substitute for this dish.

> 1 pound large white mushrooms, about 3 inches in diameter
> 2 tablespoons red wine vinegar

For Frying
> 1/2-cup virgin olive oil
> 4 tablespoons all purpose flour
> Salt, pepper

Rinse mushrooms quickly with cool water and pat dry with paper towel. Remove stem and discard. Slice each cap to make two round pieces of equal thickness. Mix flour, salt, and pepper, and toss mushroom slices in flour mixture.

In a large frying pan, heat oil. Fry both sides until golden brown. Oil should be hot. Remove from frying pan and place on paper towel to drain excess oil.

Sprinkle with vinegar and serve hot or cold.

Preparation time: 15 min.   Cooking time: 20–25 min.   Yield: 20–24 pieces

Fried Cheese
(Tyri Saganaki)

This very popular meze dish is associated with kefi and happiness. It is served at almost every Greek church festival across the US with the traditional expression "Opa!" Perfect meze for ouzo.

>   1/2-pound cheese (Kefalograviera or Kasseri)
>   Pepper
>   3 tablespoons extra virgin olive oil
>   3 tablespoons ouzo
>   1 lemon, cut in quarters

Cut cheese into 1/2-inch-thick (about 3 x 3-inch squares or 3 inch round) slices.

Mix flour with pepper and toss with cheese. In a small skillet, heat 1 tablespoon of oil. Fry both sides until golden. Oil should be hot. It takes about one minute per side, if skillet is hot.

Remove skillet from heat, add 1 tablespoon ouzo, and light with a match. This is when you make the announcement, "Opa!"

Add 1 tablespoon of oil every time you fry a new piece of cheese.

Serve hot with a lemon quarter.

Tip: To give a savory flavor to this dish, add a pinch of ground mastiha (mastich) of chiou to the flour mixture. You will find mastiha at Greek delicatessens.

Preparation time: 5 min.   Cooking time: 6–8 min.   Yield: 3 slices

Eggplants with Yogurt Sauce
(Melitzanes Tiganites me Tzatziki)

This is a very nice dish when served with Tzatziki sauce. You will need young, fresh eggplants. Try it. You will like this unique combination.

- 2 pounds eggplants
- 1 teaspoon dry yeast
- 1 cup lukewarm water
- 3 tablespoons flour
- Salt, pepper

For Frying
- 1 cup virgin olive oil
- 3 tablespoons flour
- Tzatziki sauce, see page 90

Wash eggplants and slice into 1/2-3/4-inch-thick round slices. Discard stems. Sprinkle with plenty of salt and set aside in a bowl for 30 minutes to drain.

In the meantime, in a medium mixing bowl dissolve yeast in a cup of lukewarm water. Add flour to make a runny batter (like pancake batter). Set aside for 20 minutes.

Wash eggplant with plenty of water to remove salt and squeeze to remove all water. Dry with paper towels. Toss eggplant slices in flour.

In a frying pan, heat half of the oil. Mix batter and dip the floured eggplants into batter. Fry on both sides until golden.

Place on paper towel to remove excess oil. Serve plain or with Tzatziki sauce.

Tip: Select young and shiny eggplants. Because eggplant is porous like a sponge, it will absorb oil when fried. After eggplant is salted for 30 minutes, wash with water and squeeze to remove all water and eliminate air pockets to reduce the amount of oil absorbed.

Preparation time: 40 min.   Cooking time: 25 min.   Yield: 16–20 slices

Fried Peppers
(Piperies Tiganites)

Another excellent meze for ouzo. The sweet flavors of the peppers are accentuated when prepared with good olive oil.

> 1 pound sweet red and yellow peppers
> Salt, pepper

For Frying
> 1/2 cup virgin olive oil
> 2 tablespoons red wine vinegar

Wash peppers and dry with paper towel.

Make a very small slit on the side and remove seeds without removing the stem or splitting the pepper (optional). Set aside to dry.

In a large frying pan, heat oil and fry peppers until golden brown.

Remove from heat and place on paper towel to remove excess oil.

Arrange on a platter, sprinkle with vinegar, add salt and pepper, and serve hot or cold.

Tip: Green peppers, in general, are not suitable for this dish.

Preparation time: 30 min.   Cooking time: 15 min.   Yield: 20–30 pieces

Pumpkin Fritters
(Kokkino Kolokithi Tiganito)

In the fall when pumpkin is plentiful, this recipe will provide something distinctive for your family and guests. It is easy to make and very delicious.

    2-3 pounds pumpkin
    4 cups milk
    1 cup flour
    Salt, pepper
    1 teaspoon ground cloves
    2 eggs, beaten
    1/2-stick (2 ounces) butter, melted
    3/4-cup bread crumbs, plain

For Frying
    1 cup virgin olive oil

Peel skin off pumpkin. Slice pumpkin into 3/8-1/2-inch-thick slices. Discard skin and seeds

In a medium saucepan, heat milk and add pumpkin. Cover and simmer for about 7 minutes or until pumpkin is soft, but not cooked. Remove, drain, and pat dry with paper towel; cut pieces into 2-3- inch squares and set aside.

In a small mixing bowl combine flour, salt, pepper, and cloves. In another bowl, combine eggs with butter.

In a frying pan, heat oil. Toss pumpkin pieces in flour mixture, then dip into egg mixture, and lastly toss into bread crumbs. Fry on both sides until golden and crispy.

Serve hot.

Preparation time: 20 min.   Cooking time: 20 min.   Yield: 4–6 servings

Fried Artichokes
(Anginares Tiganites)

Artichokes are best prepared and served during springtime. Fried artichokes offer a slightly different way of enjoying this healthy vegetable.

> 5 large artichokes or 6 small
> 1 lemon
> Salt
> 2 tablespoons red wine vinegar

For Frying
> 3 tablespoons flour
> Pepper
> 1/2-cup virgin olive oil

Cut one lemon in half. Pour 2 tablespoons of salt into a small dish and set aside. You will need salt and lemon to rub the artichokes after peeling to prevent oxidation.

To clean the artichokes start by breaking off the outer 4-5 layers of leaves, bending leaves towards the stem, away from you. Cut the tip end of the leaves to expose the center purple choke of the artichoke. Using a small teaspoon, remove the purple choke. Dip lemon into salt and rub exposed center of the artichoke.

With a sharp paring knife, peel off the hard part of the body until most of the green is removed and the white flesh of the artichoke appears. Again, rub with salt and lemon. Now, cut the top of the stem to about 1 inch long and peel off all hard parts until the white flesh appears. Rub again with salt and lemon, and set aside. Continue cleaning the rest of the artichokes.

Slice large artichokes into 6 wedges and small ones into 4 wedges. Wash with water and drain.

Mix flour, pepper, and salt, and toss artichokes in this mixture.

In a medium frying pan, heat the oil and fry artichokes until golden. Remove and set on paper towel to drain excess oil. Sprinkle with vinegar and serve hot or cold.

Tip: You may substitute corn meal for the flour. If you are watching your calories, you can also fry artichokes plain, eliminating the breading.

Preparation time: 40 min.   Cooking time: 20 min.   Yield: 20–30 pieces

Fried Cauliflower Florets
(Kounoupidi Tiganito)

This tasty dish will win over even those who do not particularly like cauliflower.

  2 pounds (1 medium) cauliflower head
  2 tablespoons flour
  1 egg, beaten
  3/4-cup flour
  Salt, pepper
  3–4 tablespoons Kefalograviera cheese, grated

For Frying
  1 cup virgin olive oil
  1/2 cup plain, dry, bread crumbs

Wash cauliflower and separate into small florets.

In a medium saucepan, add florets, cover with water, and bring to a boil. Cover and simmer for about 10 minutes or until florets are soft. Remove from heat; drain. Let florets cool.

In a small mixing bowl add 2 tablespoons flour and mix with 1/2-cup water. Add egg and mix well to make a thin and runny batter (like pancake batter).

Place remaining flour into a plastic bag. Add salt and pepper. Add florets and shake bag to coat with flour.

Wipe saucepan clean, add oil, and heat.

Dip floured florets into batter and then toss in bread crumbs. Fry in oil until golden. Remove from heat and place on paper towel to remove excess oil.

Place on a serving platter, sprinkle with cheese, and serve hot.

Tip: You can substitute Parmesan cheese for Kefalograviera cheese.

Preparation time: 30 min.   Cooking time: 20 min.   Yield: 20–30 florets

Grilled Mixed Vegetables
(*Lahanika sta Karvouna*)

This is an ideal dish for vegetarians and for those who like foods with a smoky flavor.

   1 medium eggplant
   Salt
   1 yellow zucchini
   1 green zucchini
   1 red bell pepper
   6–8 spears asparagus
   1 red onion
   4-5 broccoli florets
   3 tablespoons grated Kefalotyri cheese

Marinade
   1/4-cup extra virgin olive oil
   1/4-cup red wine vinegar
   3 garlic cloves, crushed and chopped
   1 tablespoon dry oregano
   Pepper

Wash vegetables. Slice eggplant into 1/2-3/4-inch-thick round slices. Sprinkle with plenty of salt, place into a bowl and set aside for 30 minutes to drain.

Slice yellow and green zucchini into 1/2-inch-thick round slices and set aside. Core bell pepper, discard stem and seeds, and cut into 4–5 slices. Peel onion and slice into 1/2-inch-thick rounds.

In a large mixing bowl, mix oil, vinegar, garlic, oregano, salt, and pepper.

Wash eggplant thoroughly to remove salt. Place all vegetables in bowl with marinade, mix well, and set aside for 30 minutes.

In the meantime, heat gas grill to high setting. Grill vegetables about 5–7 minutes on each side, brushing with marinade. Remove from grill, place on a serving dish. Sprinkle with salt, pepper, and cheese. Serve hot or cold.

Tip: Brushing the vegetables with marinade generates smoke that gives the vegetables a subtle smoky flavor.

Preparation time: 50 min.   Cooking time: 15 min.   Yield: 6 servings

Fried Mussels
(Mydia Tiganita)

I learned this recipe from my father-in-law. This is an excellent way to serve mussels as an appetizer.

    40 large mussels
    3 tablespoons flour
    1/2-cup beer
    1 lemon, cut in wedges
    Decorative toothpicks

For Frying
    1/2-cup flour
    Salt, pepper
    1/2-cup virgin olive oil
    Skordalia sauce, see page 89

Wash and scrape mussels. With a knife remove meat and set aside to drain. Discard shells.

In a small mixing bowl add 3 tablespoons flour and slowly mix with beer to make a thick but runny batter.

Mix 1/2 cup flour with salt and pepper and toss mussels in the mixture. Set aside.

In a frying pan, heat oil. Dip mussels in batter and fry until golden brown and crispy.

Remove from heat, sprinkle with salt and pepper.

Serve with lemon wedges and Skordalia sauce. Use toothpicks for serving.

Preparation time: 30 min.   Cooking time: 20 min.   Yield: 40 bite-sized pieces

Baby Beef Liver with Oregano
(Sikoti Riganato)

I know many people do not like liver, but this recipe will change their minds. The flavors of the olive oil, oregano, and dill make this a unique dish that will please everyone.

    1 pound baby beef liver
    1/2-cup flour
    1/2-cup virgin olive oil
    Salt, pepper
    2 tablespoons dry oregano
    3 scallions, chopped
    3 tablespoons fresh dill, snipped
    1 lemon, juice of

Remove veins and skin from the liver and cut into 3/4-inch cubes.

In a plastic bag, pour flour and add liver. Shake bag well to coat liver cubes with flour. Remove from bag and dust off any excess flour.

Heat oil in a medium saucepan.

Place half of the liver cubes into the frying pan and fry for about 3 minutes, turning them frequently with a perforated spoon.

Remove from heat and place on paper towel to remove any excess oil.

Fry the remaining half of the liver cubes. Place on a paper towel to remove excess oil.

Place all liver cubes onto a platter. Sprinkle with salt, pepper, and oregano.

Add scallions, dill, and lemon juice, and serve hot or cold.

Preparation time: 25 min.   Cooking time: 10 min   Yield: 50–55 bite-sized pieces

Vegetarian Stuffed Grapevine Leaves with Rice (*Dolmadakia*)

This is vegetarian Dolmadakia. This famous Greek dish combines the flavors of grapevine leaves with fresh herbs. It takes a bit longer to prepare, but the taste is worth the extra effort.

> 1 jar (net weight, drained, 16 ounces) grapevine leaves
> 1 pound onions (about 3 medium), grated
> 1 cup long grain rice, washed and drained
> 1/2-cup fresh parsley, finely chopped
> 1/4-cup fresh mint, finely chopped
> 1/4-cup fresh dill, chopped
> 1 lemon, juice of
> 1 tablespoon sugar
> Salt, pepper
> 1 cup extra virgin olive oil
> 2 lemons, cut into wedges

Open jar and remove half of the leaves. In a medium saucepan, bring 5–6 cups water to a boil. Gently unroll leaves, place into boiling water, and simmer for 15 minutes. Do not overcook leaves, as you want them to be soft, but firm when you stuff them. Remove leaves with perforated spoon and place into a colander to drain and cool.

In the same saucepan, bring 3 cups water to a boil and blanch onions for 1 minute. Drain and set aside. Wipe the saucepan clean. Line the bottom of the saucepan with 2–3 grapevine leaves.

In a large mixing bowl, combine onions, rice, parsley, mint, dill, lemon juice, sugar, salt, pepper, and one-fourth of the oil.

To stuff the leaves, place a leaf on a flat surface, stem end towards you, rough side up, and remove excess stem. Place 1/2 tablespoon of rice mixture at the stem end of the leaf.

Fold stem end towards rice mixture just half a turn. Then fold left side towards the right and the right side on top of the left. Roll loosely to allow room for the rice to expand. This way the leaves will not burst open when rice is cooked. Roll leaf to the end.

In a saucepan, arrange rolled leaves in rows close to each other with fold side down. When finished with one layer, start a new one.

Slowly add water to cover the leaves. Add the remaining oil. Place a plate on top of the stuffed leaves to keep them tight in the saucepan. Cover and simmer on low heat for about 60 minutes or until the rice is cooked and the juices are absorbed. Remove from heat and let cool.

On a serving platter, arrange stuffed leaves, garnishing with lemon wedges. Discard juices. Serve plain or with plain yogurt.

Tip: Grapevine leaves should be tender. If leaves are large, split them in half and remove hard stem. Dolmadakia should be about the size of your thumb. Grapevine leaves are available in almost every supermarket in the gourmet section. You will need to squeeze the leaves in order to remove them from the jar.

Preparation time: 1 hr. 30 min.   Cooking time: 1 hr.   Yield: 30–40 Dolmadakia

Beans with Tomato Sauce (*Gigantes Plaki*)

For those who like beans, this dish will not disappoint. The ingredients are combined and slowly cooked in the oven to give this dish a sweet flavor. Excellent as an appetizer for all occasions.

> 1 pound gigantes (giant beans) or large lima beans
> 1/2-cup extra virgin olive oil
> 2 medium onions, finely chopped
> 2 garlic cloves, finely chopped
> 1 can (14 ounces) tomato puree
> 1/2-cup water
> 2-3 bay leaves
> 1 teaspoon sugar
> 1/2-cup parsley, finely chopped
> 2 tablespoons dry oregano
> Salt, pepper
> 2 tomatoes, thinly sliced into rounds

In a medium bowl, add water and soak beans for 12 hours. Drain and discard water. In a medium saucepan, add beans and 5 cups water; boil for 15 minutes. Drain and discard water. Replace water and bring to another boil. Add beans, cover, and simmer until soft (about 90 minutes).

In the meantime, in a large skillet, heat one quarter of the oil and sauté onions and garlic. Oil should not be very hot to avoid burning the garlic.

Add tomato puree, water, bay leaves, sugar, parsley, half of the oregano, mix, then cover and simmer for 5 minutes, mixing occasionally. Remove from heat and set aside.

When beans are soft, drain and discard water. Arrange beans in a 12 x 9 x 2-inch baking dish; add tomato sauce, salt, pepper, remaining oil, and mix with beans.

Arrange the tomato slices over the beans, cover with aluminum foil; bake at 375° F for 45–50 minutes. Remove aluminum foil and bake for additional 15 minutes, or until juices are absorbed. Beans should be left with very little juice. Remove from oven, sprinkle with remaining oregano, and serve hot or cold.

Preparation time: 20 min.   Cooking time: 2 hrs. 40 min.   Yield: 6 servings

Eggplant Dip
(Melitzanosalata)

This is my mother-in-law's recipe. I believe she makes the best Melitzanosalata. Grilling the eggplants on an open flame gives this dip a unique smoky flavor.

3 pounds eggplants (about 3 large)
1 red bell pepper
1/4-cup extra virgin olive oil
3 tablespoons lemon juice
Salt, pepper
5 black olives, pitted, sliced
1 tablespoon fresh parsley, chopped
Pita bread, wedges

Wash bell pepper and eggplants. With a sharp knife, score each eggplant. This will prevent them from bursting when grilled. Place eggplants and bell pepper on hot flaming gas or charcoal grill and cook for about 30-45 minutes or until soft and slightly charred. Turn over to cook on all sides. Remove eggplants and let cool. Remove bell pepper, place into a paper bag, seal, and let cool. In a medium mixing bowl combine oil and lemon juice.

Remove and discard skin, stem, and all visible seeds from each eggplant. Preserve as much of the pulp as possible. Place pulp into a medium-sized sieve, over a bowl. Using a small coffee saucer, squeeze pulp to drain all juices. Preserve juice and set aside. Place pulp on chopping block and chop well with a large knife.

Add chopped eggplant pulp to the lemon and oil mixture and mix well. The pulp will turn a lighter color. Remove skin and seeds from bell pepper and discard. Chop three-quarters of the bell pepper into small cubes and add to eggplant mixture. Add 2 tablespoons of the preserved eggplant juice, salt, pepper, and mix. This will enhance the flavor. Discard remaining juices.

Add more lemon to taste. Mix all ingredients well and place on a serving dish. Julienne remaining bell pepper and use peppers and olive slices to garnish the dish. Sprinkle with parsley and serve.

Tip: Eggplants should be young and firm or they will have too many seeds and could be bitter. In the absence of a gas or charcoal grill, broil eggplants and bell pepper in the oven. Serve with pita bread wedges. Cover and store in refrigerator for one week.

Preparation time: 40 min.   Cooking time: 40 min.   Yield: 1–2 cups

Meat Pie
(*Kreatopita*)

Here is another tasty dish using filo pastry filled with ground lamb. Very good as an appetizer or as a main course.

    1 pound ground lamb
    1 pound filo pastry #4 or #7
    1/2 cup extra virgin olive oil
    1 cup scallions, finely chopped
    2 medium onions, finely chopped
    4 tablespoons tomato puree
    1/2 cup water
    Salt, pepper
    2 eggs
    1/4 cup light cream
    6 ounces fresh Myzithra cheese
    1/4 cup dill, finely chopped
    1/2 cup dry breadcrumbs

Remove filo pastry from refrigerator and keep at room temperature for 6 hours. If frozen, thaw at room temperature for 12 hours. Keep in original packaging.

In a large saucepan, heat 1/4 of the oil and sauté scallions and onions. Add ground lamb and brown, mixing frequently.

Add tomato puree, water, salt, and pepper. Cover and simmer over low heat until lamb is tender. Remove from heat and let cool for 5 minutes.

Beat eggs, mix with light cream, and pour into lamb mixture. Add cheese, dill, and mix. Now add bread crumbs to absorb all juices. Mixture should be thick and without juices.

Cut two 6-inch-wide filo sheets from the original packaging. Remove cut pastry from its packaging and unroll. With a kitchen towel, cover pastry to prevent from drying up and becoming brittle. Grease a 12 x 9 x 2-inch baking pan. Seal and store remaining filo in the refrigerator or the freezer.

Preheat oven to 375° F.

On a flat surface, place two #4 or one #7 filo pastry sheets with narrow side towards you. Brush with plenty of oil. Place two full tablespoons of meat mixture onto the pastry sheet at the end towards you. Lift left hand side and roll towards the right hand until it covers half of the lamb mixture. Lift right hand side and roll towards the left hand to meet the edge of the left hand side. Roll pastry to end. This will make a cylindrical-shape pastry.

Place in baking pan with open end down and brush with plenty of oil. Repeat until all meat mixture is used.

Bake for 40–45 minutes or until pastry is golden. Serve hot.

Tip: You can substitute ricotta cheese for Myzithra.

Preparation time: 50 min.   Cooking time: 45 min.   Yield: 16–20 individual puffs

Fried Potatoes with Eggs
(Tiganites Patates me Avga)

A quick recipe that is easy to prepare and children love it.

>2 medium potatoes
>1/4 cup virgin olive oil
>2 eggs
>3 tablespoons milk
>2-3 sun-dried tomatoes, chopped
>2-3 tablespoons Kefalotyri cheese, grated
>Salt, pepper
>1 tablespoon ketchup
>2-3 fresh basil leaves, coarsely chopped

Peel, wash, and cut potatoes into French fries. In a medium frying pan, heat oil and fry potatoes. Remove from heat, retain 2 tablespoons of the oil, and discard the rest.

In a small mixing bowl beat eggs. Add milk, tomatoes, cheese, salt, and pepper. Heat the potatoes in the frying pan in the remaining 2 tablespoons of oil. Pour egg mixture over the potatoes, mix, and cook until eggs are done on one side. Remove from heat, flip over, and cook on the other side.

Place into a serving dish, garnish with ketchup and basil. Serve hot or cold.

Tip: You can substitute Parmesan or sharp cheddar cheese for the Kefalotyri cheese.

Preparation time: 15 min.   Cooking time: 20 min.   Yield: 2 servings

Eggs with Tomatoes
(Strapatsada)

The summertime, when fresh, ripe tomatoes are plentiful, is the best time to make this dish. Can be served as breakfast, light lunch, or as a meze with ouzo.

    2 eggs
    2 large ripe tomatoes or 1 (10 ounces) can chopped tomatoes
    2 tablespoons extra virgin olive oil
    1/2 yellow bell pepper, coarsely chopped
    1/2 teaspoon sugar
    3 tablespoons milk
    1/4 cup Feta cheese, crumbled
    Salt, pepper
    3-4 basil leaves, coarsely chopped

Peel tomatoes, remove seeds, and chop into 1/2-inch cubes.

In a medium frying pan heat oil and sauté bell pepper until soft, mixing frequently. Add tomatoes and sugar. Simmer until most of the juices are absorbed.

In the meantime, in a small mixing bowl beat eggs. Add milk, Feta cheese, salt, and pepper. Mix to blend. Pour egg mixture into tomato sauce and simmer uncovered over a low heat. As eggs are setting up, cut with a fork to keep mixture fluffy.

Remove from heat and place into a serving dish. Sprinkle with basil leaves. Serve hot or cold.

Tip: Skip milk if you cook eggs sunny-side up. To make tomato peeling easier, using a sharp knife, make two small slits in the tomatoes, and then blanch for 1 minute in boiling water. Remove and place into ice water and then remove skin.

Preparation time: 15 min.   Cooking time: 15 min.   Yield: 2 servings

Shrimp with Lemon Sauce
(Garides me Lemoni)

This is a simple, but very delicious dish that blends the subtle flavors of succulent shrimp, lemon, and aromatic extra virgin olive oil.

   1 pound (20-25) shrimp
   2 tablespoons parsley, finely chopped
   1/2 cup Ladolemono sauce, see page 92

In a medium saucepan, add 4 cups of water and bring to a boil.

Add shrimp, and when water starts boiling again remove from hot water and place into ice cold water. When cooled, remove shells, devein, and place into a serving dish.

In the meantime, prepare Ladolemono sauce.

Pour sauce over shrimp. Sprinkle with parsley and serve.

Preparation time: 15 min.   Cooking time: 2 min.   Yield: 4 servings

Fried Squid
(Kalamaria Tiganita)

This is a very popular dish on the island of Mytilini. Use fresh squid for the best result. Can be served as an appetizer or as a main entrée.

2 pounds squid (5-6 squid about 6 inches long)

For Frying
1/2-cup flour
Corn meal
Hot paprika (optional)
Salt, pepper
1 lemon, cut in wedges
1 cup virgin olive oil
1 cup Skordalia sauce, see page 89

Prepare squid by removing and discarding backbone and ink. Remove and discard head, but keep tentacles. Wash main body and tentacles with cold water, drain, and set aside.

In a medium mixing bowl combine flour, corn meal, paprika, salt and pepper.

On a chopping board, slice squid into 1/2-inch rings. Ensure squid is well-drained. Toss pieces in flour mixture. Dust off excess flour.

In a deep frying pan, heat oil and fry squid until golden.

Remove with large perforated spoon and place on paper towel to remove excess oil.

Serve hot with lemon wedges and Skordalia sauce.

Tip: Select a squid which has some thickness to its body. This way the squid will not dry up and will not be chewy when cooked. Cover your frying pan with a splatter screen to prevent oil from splashing.

Preparation time: 30 min.   Cooking time: 10 min.   Yield: 4 servings

Grilled Shrimp
(Garides stin Skara)

This delicious dish will please even the most finicky eater. It is best when cooked on the grill. Your family and friends will rave about this one.

    1 pound (20–25) shrimp, shelled and deveined
    2 tablespoons extra virgin olive oil
    1/4-cup white wine
    1/2-teaspoon dry mustard
    Salt, pepper
    10–12 wooden skewers, 10 inches long
    1/2-cup Ladolemono sauce, see page 92

Soak wooden skewers in water for 30 minutes. Prepare Ladolemono sauce, and refrigerate. In a medium mixing bowl, combine oil, wine, mustard; marinate shrimp in this liquid for 2–3 hours. Arrange 4–5 shrimp on double skewer.

Grill over hot charcoal or gas grill for about 3-4 minutes per side. Brush frequently with marinade. Remove from heat and dip into Ladolemono sauce. Serve hot or cold.

Tip: When grilling, use two skewers to support the shrimp. This way shrimp will not spin when you turn them over.

Preparation time: 10 min.   Cooking time: 10 min.   Yield: 4-5 servings

Caviar Croquettes
(Taramakeftedes)

A very unusual dish prepared with bread, caviar, and herbs. Croquettes are formed into patties and fried in olive oil. Delicious! An excellent meze for ouzo.

   10 ounces (10 slices) white bread
   3 tablespoons Tarama (caviar)
   1/4 cup parsley, finely chopped
   2 tablespoons dill, chopped
   2 tablespoons virgin olive oil
   1 small onion, grated
   Pepper
   1 tablespoon fresh oregano, chopped
   2-3 tablespoons bread crumbs

For Frying
   1/2-cup virgin olive oil
   1/2-cup flour

Remove crust from bread and soak in water for a couple of minutes. Squeeze water out and discard.

In a medium mixing bowl, combine bread, Tarama, parsley, dill, oil, onion, pepper, and oregano. If mixture is too soft, add 2-3 tablespoons bread crumbs.

Using a tablespoon, make balls the size of a large walnut. With your fingers, gently squeeze and flatten balls into small patties. Toss in flour and set aside.

In a large frying pan, heat half of the oil and fry croquettes on both sides until golden. Fry all croquettes, adding more oil as required. Remove and place on paper towel to drain excess oil.

Serve hot or cold.

Tip: Tarama (caviar) is salty; therefore, there is no need for salt. Dip spoon in water to keep the mixture from sticking to the spoon. Bread should be 2-3 days old or lightly toasted.

Preparation time: 30 min.   Cooking time: 20 min.   Yield: 16–18 croquettes

Cheese Puff Bites with Homemade Pastry (Tyropitakia)

Salty cheese wrapped in sweet and rich homemade dough makes this dish unique and pleasing. Fantastic cheese puffs for any occasion.

Dough
>	10 ounces all-purpose flour
>	1 teaspoon salt
>	4 ounces butter, room temperature
>	2 egg yolks
>	1/4-cup ice cold water

Filling
>	5 ounces Kefalotyri cheese, grated
>	1 egg yolk
>	3 tablespoons milk
>	3 tablespoons parsley, finely chopped
>	Pepper
>
>	2 tablespoons extra virgin olive oil
>	1 egg yolk beaten in 2 tablespoons water

To Make Dough
Combine flour and salt; sift twice. Add flour to food processor. Cut butter into 1/2 -inch cubes; add to flour. Add egg yolks; mix at low speed for approximately 5 minutes.

When butter is mixed with flour, add water, one tablespoon at a time. When dough becomes a ball, stop mixing and DO NOT add any more water. Remove from bowl, shape into a ball, wrap with plastic wrap and refrigerate for 30 minutes. Can keep refrigerated 2–3 days; or frozen for a longer period of time.

To Prepare Filling
In a medium mixing bowl, combine cheese, egg yolk, milk, parsley and pepper; set aside. Remove dough from refrigerator and discard plastic wrap. Roll dough on lightly floured flat surface to about a 1/8-inch-thick sheet.

Cut dough into 3-inch-diameter rounds (large glass or coffee cup is ideal for this). Fill each round with 1 flat teaspoon of cheese filling. Wet dough edges with water and fold over to form a half-circle and pinch ends to seal. Preheat oven to 350° F.

Arrange puffs on greased baking pan. Brush with egg yolk. Bake for 35-40 minutes or until golden. Serve hot or cold.

Preparation time: 1 hr. 10 min.   Cooking time: 35-40 min.   Yield: 24–28

Alternate Filling

  5 ounces ground beef
  2-3 tablespoons virgin olive oil
  1 small onion, finely chopped
  1 tablespoon tomato paste, dissolved in 3 tablespoons water
  3 tablespoons dill, finely chopped
  Few drops of hot sauce (optional)
  1/2 teaspoon sugar
  Salt, pepper
  1-2 tablespoons plain dry bread crumbs
  1 egg yolk, beaten in 2 tablespoons water

In medium saucepan, sauté onion in oil. Add meat; brown 5 minutes; mix frequently.

Add tomato paste, dill, water, hot sauce, sugar, salt, pepper; simmer, covered until meat is soft and juices absorbed. Add bread crumbs; mix well (bread crumbs will absorb juices). Let cool. Follow instructions for filling the dough.

Fried Zucchini
(Kolokithia Tiganita)

Fried zucchini is easy to make and it is great as an appetizer or as side dish. Use green or yellow zucchini.

> 2 medium (24 ounces) green or yellow zucchini
> 2 tablespoons flour
> 1/2-teaspoon baking powder
> Salt, pepper

For Frying
> 1/2 cup virgin olive oil
> 1/2 cup flour
> 2 tablespoons red wine vinegar
> 2 ounces Kefalotyri cheese, grated

Wash zucchini and cut into rounds about 3/8 inch thick.

In a small mixing bowl combine 2 tablespoons flour, baking powder, salt, pepper, and mix well. Add 1/2-cup water; mix well to make a runny batter.

In a frying pan, heat half of the oil. Toss zucchini slices into flour, dip into batter, and fry in hot oil until golden.

Place fried zucchini on a paper towel to drain excess oil.

Sprinkle with vinegar and cheese and serve hot.

Tip: You can substitute Parmesan cheese for Kefalotyri cheese.

Preparation time: 20 min.   Cooking time: 30 min.   Yield: 45–50 slices

SALADS

A Greek salad is not just tomato and Feta cheese. Greeks love salads. There are many kinds of refreshing salads, from a simple cucumber salad with vinegar and oil to more elaborate salads, like Russian salad made with various vegetables, pickles, capers, and mayonnaise.

Salads of any kind are very important in the Greek cuisine. They are served as light lunches or as refreshing side dishes. Greeks always eat their salad with the meal; never before and never after unless they just have salad alone.

The best-known salad, made during the winter months, is made with romaine lettuce. Greeks cut their lettuce thin and add dill, scallions, arugula, and sometimes, fresh mint and parsley.

Another well-known Greek salad is made with fresh ripe tomatoes (skin always removed), Feta cheese, slices of red onion, olives, and few drops of aromatic extra virgin olive oil. When the tomatoes are ripe, the salad bowl is full of red tomato juices in which Greeks love to dip their crusty bread. Do not be surprised to see people competing for the juices.

Vegetables, like beans, string beans, zucchini, beets, and others are served as salads. Greeks love these vegetables. Vegetables are simply boiled in water and then dressed with olive oil and lemon (Ladolemono), or olive oil and vinegar (Ladoxido).

Dry vegetables, like beans, are garnished with olives, hard-boiled eggs, scallions, parsley, or red peppers and served as salad. Garnishing is left to the imagination of each cook.

Greeks love to serve green salads like dandelion, lettuces, string beans, and zucchini with grilled fish, and roasted lamb. Tomato salads are popular as summer dishes, served with fried fish, roasted or baked chicken, or just as a plain tomato salad accompanied by some good Feta cheese.

Boiled Zucchini Salad
(Kolokithakia Vrasta)

This salad makes an excellent accompaniment to any fish dish. It is very light and the extra virgin olive oil adds a nice subtle flavor to this simple salad.

> 2 pounds small green zucchini (about 1 inch thick)
> 2 tablespoons extra virgin olive oil
> 1 lemon, juice of
> Salt, pepper

In a medium saucepan, boil water, add zucchini; boil until soft (about 5 minutes). Zucchini should not be overcooked.

Remove from heat, drain, and place into ice water to cool down quickly and stop the cooking process. When cooled, remove from ice water and drain.

Cut ends and discard. Slice into 1/2 inch rounds and place into a colander for few minutes to drain excess water. Arrange zucchini slices on a serving dish.

In a small mixing bowl beat lemon with oil, and sprinkle over the zucchini.

Add salt and pepper. Serve hot or cold.

Preparation time: 10 min.   Cooking time: 5–8 min.   Yield: 4 servings

Red Beet Salad
(*Pantzaria Vrasta*)

This is an excellent salad to serve with fried fish. It is simple, but very delicious, especially when served with Skordalia sauce.

3 pounds red beets (bulbs and leaves)
4–5 tablespoons extra virgin olive oil
2–3 tablespoons red wine vinegar
1 medium garlic clove, crushed
Salt
Skordalia sauce (optional), see page 89

Separate bulbs from leaves. Cut roots, wash and place into a large-sized saucepan with boiling water; cover and simmer.

In the meantime, select and keep the tender leaves and stems. Cut into 3- or 4-inch long pieces, wash well and set aside.

When bulbs are almost done (takes about 10–12 minutes) add stems and leaves and cook for few minutes until stems are soft. Remove from heat, drain, and set aside to cool. Separate bulbs from leaves and stems. Remove skin from the bulbs and slice into 1/4-inch-thick rounds. Arrange sliced bulbs in the middle of a serving platter with leaves and stems around the bulbs.

In a small mixing bowl blend oil and vinegar. Add garlic, mix, and pour over beets. If you use Skordalia sauce, omit the crushed garlic from the oil-vinegar mixture.

Sprinkle with salt and serve warm or cold. Serve with fried fish or fried bakala.

Tip: Wash beets well to remove all sand or soil. Purchase beets with about 2–3-inch bulbs and shiny leaves.

Preparation time: 15 min.   Cooking time: 20 min.   Yield: 6 servings

String Bean Salad
(Fasolakia Vrasta Salata)

A good, refreshing salad. Just-boiled fresh beans with lemon and extra virgin olive oil make this an ideal salad to serve with fried fish.

    1 pound fresh green beans
    3 tablespoons extra virgin olive oil
    1 lemon, juice of
    1 garlic clove, crushed
    Salt
    Skordalia sauce (optional), see page 89

Select green, shiny, tender beans. Remove strings and wash.

Cut beans into 3- or 4-inch long pieces and cook in boiling water until soft. Remove from heat, drain, and place in ice water. This will stop the cooking process.

Drain well and place onto a serving platter.

Blend oil with lemon juice, add garlic, and pour over string beans. If you are using Skordalia sauce, omit the crushed garlic from the oil-lemon mixture.

Sprinkle with salt and serve.

Tip: Best prepared using black-eyed beans, string beans, or the yard-long round beans.

Preparation time: 25 min.   Cooking time: 30 min.   Yield: 4 servings

Black-Eyed Bean Salad
(Fasolia Mavromitika)

This is another simple salad for those who like dry beans. The unique flavor comes from the extra virgin olive oil and the sweetness of the beans.

    8 ounces dry black-eyed beans
    1 small onion, peeled, whole
    Salt
    4 tablespoons extra virgin olive oil
    2–3 scallions, finely chopped
    Pepper

Wash beans in cold water. In a medium saucepan, add beans, cover with water, and boil for 10 minutes. Remove from heat and drain. Add new, fresh water to cover beans and return to heat.

Add onion and simmer until beans are soft. When soft add salt and simmer another 5 minutes. Remove from heat, drain, and discard onion.

Place into a serving bowl, add oil, and mix with beans. Sprinkle with scallions and pepper. Serve hot or cold.

Tip: As a substitute for dry beans, use frozen or canned. If you are using canned beans, first wash with water, and drain. Boil canned beans for 10 minutes only.

Preparation time: 10 min.   Cooking time: 40 min.   Yield: 4 servings

Cauliflower Salad
(Kounoupidi Vrasto Salata)

Boiling cauliflower, just enough to soften it, and then mixing it with an oil and lemon dressing is another way to enjoy this vegetable. An excellent salad to serve with baked meats or baked poultry.

2 pounds (1 medium) cauliflower head
4 tablespoons extra virgin olive oil
1 lemon, juice of
Salt, pepper

Wash cauliflower and separate it into small florets.

In a medium saucepan, add florets, cover with water, and bring to a boil. Cover and simmer for about 10 minutes or until florets are soft; do not overcook. Remove from heat, place in cold water to stop the cooking process; drain well.

Place into a serving bowl. Blend oil and lemon juice and pour over cauliflower.

Sprinkle with salt and pepper. Serve hot or cold.

Preparation time: 10 min.   Cooking time: 10 min.   Yield: 4 servings

Carrot Salad
(Karotosalata)

This is an easy salad to make and if you like carrots; you will enjoy the sweetness of the carrots, which are enhanced by the aroma of the extra virgin olive oil. Great salad with fish or poultry.

  2 pounds (4–5 large) carrots
  1 lemon, juice of
  2–3 tablespoons extra virgin olive oil
  Salt, pepper

Peel and wash carrots. Grate carrots and place into a medium mixing bowl. Discard any water.

Combine lemon juice with oil and mix with carrots. Sprinkle with salt and pepper. Serve cold.

Tip: Serve immediately after adding salt to prevent carrots from becoming soft and watery.

Preparation time: 15 min.   Yield: 4 servings

Dandelion Salad
(Radikia Vrasta)

Wild dandelions make the best salad. However, if wild dandelions are not readily available, you can purchase farm-grown dandelions, or collards. Very good with grilled fish or grilled lamb chops.

  2 pounds (1–2 bunches) dandelions
  1 lemon, juice of
  3–4 tablespoons extra virgin olive oil
  Salt and pepper

Remove and discard roots and damaged leaves. Cut dandelion leaves into 2–3-inch long pieces and wash in cold water 2–3 times.

Place dandelions into a large saucepan, cover with water, and bring to a boil. Cover and simmer over a low heat for about 15–20 minutes until dandelions are soft. Do not overcook. Remove from heat and drain in a colander. Squeeze lightly to remove excess water and then place onto a serving dish.

In a small mixing bowl combine lemon juice with oil and pour over dandelion greens. Sprinkle with salt and pepper. Serve hot or cold.

Tip: Dandelions need a thorough washing to remove sand or soil. Purchase dandelions with bright green leaves.

Preparation time: 20 min.   Cooking time: 15–20 min.   Yield: 4 servings

Mixed Salad
(*Lahano-Marouli kai Karoto Salata*)

Cabbage, carrots, and lettuce go well together when combined with red wine vinegar and extra virgin olive oil.

> 1 small romaine lettuce, thinly sliced
> 1 large carrot, peeled and grated
> 1 cup green cabbage, thinly sliced
> 1 cup red cabbage, thinly sliced
> 1 small red onion, thinly sliced

Salad Dressing
> 3 tablespoons red wine vinegar
> 4 tablespoons extra virgin olive oil
> Salt, pepper

Make salad dressing by combining vinegar and oil; add salt and pepper.

In a medium mixing bowl add 1 tablespoon of the oil-vinegar mixture and toss lettuce. Remove and place into a salad bowl.

Repeat for carrots, green cabbage, and red cabbage, placing each one of the vegetables next to each other on top of the lettuce. Pour remaining oil-vinegar mixture over greens and vegetables. Lastly, add red onions on top.

Serve cold.

Tip: Serve immediately after mixing in oil and vinegar, to prevent vegetables from getting soft.

Preparation time: 20 min.   Yield: 4 servings

Tomato and Cucumber Salad
(Horiatiki Salata)

Fresh vegetables and good olive oil are the secrets to this recipe. Serve with Souvlaki, with fish, or just plain. Refreshing!

    1 small red onion, thinly sliced
    Salt
    1 large ripe but firm tomato, skin removed
    1 small cucumber, peeled and thinly sliced
    6–8 green or black olives, pitted
    1 cup arugula, coarsely chopped
    2 tablespoons fresh parsley, finely chopped
    4–5 slices of green or yellow bell pepper
    2 tablespoons fresh mint, chopped
    2–3 tablespoons extra virgin olive oil
    1 tablespoon red wine vinegar
    Pepper
    4 fresh basil leaves, coarsely chopped
    1 tablespoon dry oregano

In a small mixing bowl, sprinkle onion with salt and set aside for 10 minutes.

In a large mixing bowl, slice tomato into bite-sized pieces. Add cucumber, olives, arugula, parsley, bell pepper, and mint.

Combine oil with vinegar, add salt and pepper, and mix with vegetables.

Place vegetables into a serving bowl.

Wash onion in cold water, squeeze water out, and sprinkle over salad. Garnish with basil and oregano and serve.

Tip: You may also substitute dry oregano for the fresh oregano, but omit the basil leaves.

Preparation time: 20 min.   Yield: 1–2 servings

Greek Salad
(*Domatosalata*)

A simple, but delicious salad. This is one of the most popular salads served in Greece. It is best when prepared using fresh, ripe tomatoes. Good Feta cheese is also an important ingredient. Serve as a light lunch or as a side dish.

1 large ripe but firm tomato, skin removed
4–5 slices of red onion
6–8 green or black olives, pitted
2 ounces Feta cheese, cut in small cubes
2 tablespoons fresh parsley, coarsely chopped
4–5 slices of green bell pepper
4 fresh basil leaves, coarsely chopped
3–4 anchovies (optional)
2–3 tablespoons extra virgin olive oil
1 teaspoon dry oregano
Salt, pepper

In a medium mixing bowl, slice tomato into bite-sized pieces; add onions, olives, Feta cheese, parsley, bell pepper, basil leaves, and anchovies.

Pour olive oil, toss all ingredients together, and sprinkle with oregano, salt, and pepper.

Preparation time: 10 min.   Yield: 1–2 servings

Green Salad
(Marouli Salata)

This is another salad that is very popular in Greece, especially in winter or spring, when fresh and tender romaine lettuce is available. Excellent served with grilled fish or roasted lamb. It is also very good served alone as a light lunch.

1 head romaine lettuce, thinly sliced
1 bunch scallions, coarsely chopped
2 tablespoons fresh dill, snipped
1 bunch arugula, coarsely chopped
3–4 radishes, sliced
2 tablespoons extra virgin olive oil
2 tablespoons red wine vinegar
Salt, pepper
1–2 pita breads, toasted

In a medium mixing bowl, combine lettuce, scallions, dill, arugula, and radishes. Pour oil and vinegar over the greens and toss to coat all vegetables.

Sprinkle with salt and pepper.

Place into a serving bowl and serve with toasted pita bread.

Tip: Use young, fresh lettuce, as it is the sweetest and most tender.

Preparation time: 10 min.   Yield: 2–4 servings

White Bean Salad
(Fasolia Piaz)

This is a traditional bean salad that can be served cold with dill and scallions.

    1/2 pound white beans (cannelini or Northern beans)
Salt
2 tablespoons fresh dill, finely chopped
4 tablespoons scallions, finely chopped
4 tablespoons extra virgin olive oil
2 tablespoons red wine vinegar
1 egg, hard-boiled and quartered
8–10 black olives, pitted and sliced
Pepper
1/2-teaspoon paprika

In a medium saucepan, cover beans with water and bring to a boil. Reduce heat and simmer for 20 minutes. Remove from heat, drain, and discard water.

Replace water, cover, and simmer until beans are soft.

When soft, add salt and cook an additional 10 minutes. Remove from heat and drain. Discard water. Place beans into a serving bowl. Add dill and scallions.

Combine oil and vinegar and pour over beans. Toss to coat beans with the oil mixture.

Garnish with eggs and olives. Sprinkle with pepper and paprika. Serve hot or cold.

Tip: Soak beans for 12 hours prior to cooking. You may substitute 2 (8-ounces) cans white beans for the dry beans. Wash in cold water and cook for 10 minutes only.

Preparation time: 10 min.   Cooking time: 1 hr. 30 min.   Yield: 4 servings

Russian Salad
(Rosiki Salata)

This is a very refreshing salad that combines a variety of vegetables with pickles, capers, and mayonnaise.

1 (6 ounces) can red beets, cut to 1/2-inch cubes
1 (14 ounces) can potatoes, cut to 1/2-inch cubes
1 (14 ounces) can lima beans, drained
1 (14 ounces) can, chopped mixed vegetables (peas, carrots, corn), drained
3 dill pickles, cut to 1/2-inch cubes
3 tablespoons capers
Salt, pepper
5 tablespoons mayonnaise
2 hard-boiled eggs, quartered
1/2 teaspoon paprika
1 tablespoon fresh dill, snipped

Using a colander, wash red beets with plenty of cold water. Repeat until beets do not leach red juices. If beets are not washed well, they will leach and your dish will turn red. Drain and place on paper towels to dry well.

In a large mixing bowl, mix beets, potatoes, lima beans, mixed vegetables, pickles, and capers. Add salt and pepper. Fold mayonnaise into the mixture.

Place into a serving bowl. Drain any excess juices.

Sprinkle eggs with paprika and garnish salad dish. Sprinkle with dill before serving.

Keep refrigerated. Can be covered and stored in the refrigerator for 2 weeks

Preparation time: 30 min.   Yield: 4–6 servings

SOUPS

Soup is not what comes to mind when most people think of Greek cooking.

Greek soups are seasonal or associated with holy traditions. Soups may be hearty and comprise an entire meal, or they may be light, and served before formal meals, like chicken Avgolemono or fish soup.

Nothing compares to eating a hot, hearty bowl of bean soup with toasted crusty bread on a cold winter's day. Lentil soup is the favorite soup to eat on Good Friday.

Bean Soup
(Fasolada)

This is a traditional Greek winter dish and probably the most famous soup in Mytilini. You will find this hearty soup served in every part of Greece. It is excellent as an appetizer or as the main entree.

1/2-pound white beans, cannelini or Northern
1 tablespoon tomato paste, dissolved in 1/2-cup cold water
1/2 (14 ounces) can tomato puree
1 medium carrot, peeled and chopped
1 medium onion, chopped
1/2-cup celery stalk, chopped
1/4-cup extra virgin olive oil plus 4 tablespoons
3 garlic cloves, halved
2–3 bay leaves
1 red hot pepper, whole (optional)
Salt, pepper
4 slices crusty bread, toasted

In a medium saucepan, add enough water to cover beans and bring to a boil. Reduce heat and simmer for 20 minutes. Remove from heat, drain, and discard the water.

Again, add enough fresh water to cover the beans, add tomato paste, tomato puree, carrot, onion, celery, 1/4 cup oil, garlic, bay leaves, hot pepper (if desired), and bring to a boil.Cover, reduce heat, and simmer for 90 minutes or until beans are soft. When beans are almost cooked, add salt and pepper. Add more water if required, so that the soup is not thick.

Brush bread slices with oil and broil in oven until crispy and golden. Remove from oven and sprinkle with salt. When beans are soft, remove from heat. Discard bay leaves and hot pepper before serving. Sprinkle about 1 teaspoon of olive oil per serving and serve with bread.

Tip: Although soaking is not necessary, you can soak the beans in water for 12 hours prior to cooking. This will help beans to cook easier. Alternatively, you can use cooked beans. Use 2 (8 ounces) cans of beans. Rinse with cold water and drain. Cook for 30 minutes or until vegetables are soft and flavors are blended.

Preparation time: 20 min.   Cooking time: 2 hrs.   Yield: 4 servings

Vegetable Soup
(Hortosoupa)

This is a very good soup that can be served year-round. Your imagination is the limit when it comes to what vegetables to use. This is my wife's recipe and on occasion, she surprises me with a bowl of soup.

1 small leek (white part only), sliced
1 small green zucchini, diced
1 small yellow zucchini, diced
1 small carrot, peeled and sliced
1 celery stalk, cut into small cubes
1 small onion, coarsely chopped
6 garlic cloves, whole
1 red bell pepper, seeds removed and chopped
1 small eggplant, cut into small cubes
4 tablespoons extra virgin olive oil
2 cups chicken or beef broth
2 medium, ripe tomatoes, chopped
Salt, pepper
1/2-cup croutons
1/2-cup Kefalotyri cheese, grated

In a large frying pan, heat half of the olive oil and sauté leek, green and yellow zucchini, carrot, and celery. Remove and set aside.

Add remaining oil and sauté onion, garlic, pepper, and eggplant.

In a large saucepan, add 2 cups water, broth, and bring to a boil. Add vegetables, tomatoes, salt, and pepper; cover and simmer for 40 minutes or until vegetables are soft. Remove from heat and set aside to cool.

Place a food mill on top of another medium or large saucepan, and using a ladle, pour vegetables into food mill and pass through. When completed, discard solids left in the food mill.

Return saucepan with soup to heat, cover, and simmer for 10 minutes. Add more water if required, to achieve the consistency and texture you prefer.

Serve hot with croutons and cheese.

Tip: You can add broccoli and string beans or substitute any of the vegetables. You may also substitute Parmesan or Romano cheese for the Kefalotyri cheese.

Preparation time: 30 min.   Cooking time: 50 min.   Yield: 6–8 servings

Lentil Soup
(Fakes)

Lentil soup can be made with tomato sauce, or prepared using only vinegar, oil, and garlic. A tablespoon of aromatic extra virgin olive oil in the serving bowl adds a nice flavor to this soup.

> 1/2 pound dry lentils
> 1/4 cup extra virgin olive oil plus 4 tablespoons
> 1 medium onion, finely chopped
> 1 small celery stalk, finely chopped
> 3 garlic cloves, sliced
> 2–3 bay leaves
> Salt, pepper
> 3–4 tablespoons red wine vinegar
> 1/2-cup croutons

In a medium saucepan, add enough water to cover lentils and bring to a boil. Reduce heat and cook for 10 minutes. Remove from heat, drain, and set aside. Discard water.

Wipe saucepan clean. Heat 1/4 cup olive oil; add onion, celery, garlic, and sauté for 2-3 minutes. Add 4 cups of water, bay leaves, lentils, salt and pepper; cover and simmer over a low heat until lentils are soft, about 30 minutes.

Place into serving bowls, sprinkle with some of the oil and vinegar, add croutons. Serve hot.

Tip: If thicker soup is preferred, dissolve 1 teaspoon of flour in the vinegar and add to the soup. Bring to boil and serve.

Preparation time: 20 min.   Cooking time: 40 min.   Yield: 4–6 servings

Beef Soup with Orzo
(*Kreatosoupa me Manestra*)

This soup uses meat stock as a base. Shank meat is the best meat to use when prepa
ing this soup. Excellent flavor and not very fatty.

      2 pounds beef shanks with bone
      1 small onion, peeled and cut in half
      4–5 strings parsley
      2 medium carrots, peeled and cut in half
      1 small celery stalk, cut in half
      2 small ripe tomatoes, seeds removed
      1 small red bell pepper, seeds removed
      Salt, pepper
      1/2-cup orzo
      2 tablespoons extra virgin olive oil
      1 lemon, juice of (optional)
      1/4-cup Kefalotyri cheese, grated
      Croutons

In a medium saucepan, add enough water to cover beef shanks and bring to a boil.
Reduce heat and using a large spoon, skim foam from the pan and discard.

Add onion, parsley, carrots, celery, tomatoes, bell pepper, salt, pepper, and cover.
Simmer until meat is tender. Remove saucepan from heat and using perforated spoon,
remove meat and place onto a dish. Cover with aluminum foil and set aside. Drain and
preserve 5 cups broth and vegetables. Pass vegetables through food mill and keep
about 1 cup of the pulp. Discard remaining pulp.

Wipe saucepan clean, add broth, and bring to a boil. Supplement with water, if needed. Add orzo and pulp, then cover and simmer until orzo is soft.

Cut meat into small, bite-size pieces and add to soup. Discard bones. Simmer for 5
minutes and remove from heat. Sprinkle soup with pepper, oil, lemon, cheese,
croutons, and serve.

Tip: In place of the orzo, you may also use 1/4-cup rice or 1/2-cup of small alphabet
pasta, if you prefer.

Preparation time: 15 min.   Cooking time: 2 hrs.   Yield: 4–6 servings

Meatball Soup
(*Giouvarlakia*)

This is a hearty soup that combines meat, rice, and a lemony Avgolemono sauce. Delicious served alone or as a main course.

   1 pound ground beef or mix beef and lamb
   1/3-cup rice
   1 medium onion, grated
   2 eggs whites (keep yolks to make Avgolemono sauce)
   4 tablespoons parsley, coarsely chopped
   Salt, pepper
   1 medium potato, cut into small cubes
   1 medium carrot, peeled and cut into small cubes
   3 tablespoons extra virgin olive oil
   2 tablespoons fresh dill, chopped.
   Pinch hot paprika (optional)
   Thick Avgolemono sauce, see page 91

Rinse rice with cold water, drain, and set aside. In a small frying pan, heat oil and sauté onion until soft, but not browned. Remove from heat and set aside to cool.

In a medium mixing bowl, combine rice with ground meat. Add egg whites, parsley, salt, pepper, onion, and mix. Shape mixture into small balls (the size of a small walnut) and place into a medium saucepan. Add enough water to cover meatballs, cover, and simmer over a very low heat. When partially cooked (about 25 minutes), arrange potatoes and carrots on top of the meatballs, cover, and simmer over a very low heat until potatoes are soft.

Before removing from heat, make sure that you have about 2–3 cups of broth remaining in the pan. Add more water if needed. Use 2 cups of hot broth to make Avgolemono sauce. Pour sauce over meatballs.

To serve, place 2–3 meatballs into the center of a soup dish, add potatoes, carrots, and cover with Avgolemono sauce. Sprinkle with dill, pepper, and hot paprika if desired. Serve hot.

Tip: Use some hot paprika to spice the dish. Carrots give a nice color to this dish.

Preparation time: 30 min.   Cooking time: 40–45 min.   Yield: 24–28 meatballs

Chicken Soup with Egg-Lemon Sauce (*Kotosoupa Avgolemono*)

This is a traditional and very popular Greek soup served in nearly every Greek restaurant across the US.

1 (2-3 pounds) whole chicken
1 carrot, peeled and cut in half
1 celery stalk, cut in half
1 medium onion, peeled and cut in half
Salt and pepper
1/2-cup rice or angel hair pasta
Thin Avgolemono sauce, see page 91

Rinse chicken in cold water, drain, and remove all visible fat.

Place chicken into a large saucepan, cover with cold water. Add carrot, celery, onion, salt, and pepper; then cover and simmer until chicken is tender. Remove saucepan from heat. Remove chicken and place onto a dish. Cover with aluminum foil and set aside. Drain vegetables, but reserve 5 cups of the broth. Discard vegetables. Strain broth and skim off excess fat.

Wipe saucepan clean, add broth, and bring to a boil. Add rice, cover, and simmer over low heat until rice is soft. Remove bones and skin from chicken and discard.

Using both white and dark chicken meat, make 2 cups of diced meat (set aside remaining meat and see tip). Add to rice and simmer for 10 minutes. Remove from heat.

Use hot broth from the soup to make Avgolemono sauce. Add sauce to soup, return to heat and simmer over a low heat for 2–3 minutes.

Sprinkle with pepper and serve hot.

Tip: Cut remaining meat into small cubes, add mayonnaise, chopped celery, salt, and pepper, and make chicken salad or serve chicken meat sprinkled with a little fresh lemon juice as a side dish with the soup. If you cover your chicken salad with plastic wrap or store it in an airtight container and refrigerate it, it will keep for a couple of days.

Preparation time: 30 min.   Cooking time: 1 hr. 30 min.   Yield: 4–6 servings

Chickpea Soup
(Revythosoupa)

This soup is very popular in winter. It takes some time to cook, but it is very tasty and worth the effort.

  1/2-pound dry chickpeas, soaked 12 hours
  1 tablespoon baking soda
  1 (10 ounces) can chicken broth
  2 tablespoons tahini
  1 small onion, finely chopped
  1 small celery stalk, finely chopped
  Salt, pepper
  1/4 cup extra virgin olive oil plus 2 tablespoons
  1 small red onion, finely chopped
  2 tablespoons fresh parsley, finely chopped

In a medium mixing bowl dissolve baking soda in water and add chickpeas. Make sure there is enough water to cover chickpeas. Soak for 12 hours, drain, and rinse well. Discard water.

In a medium saucepan, bring 4 cups of water and the chicken broth to a boil. Add chickpeas, cover, and simmer for about 60 minutes. When about half cooked add tahini, onion, celery, salt, pepper, and 1/4-cup olive oil; simmer until chickpeas are soft.

When ready to serve, add red onion and parsley; pour some oil over the soup and serve hot.

Tip: Substitute dry chickpeas with 2 (8 ounces) cans chickpeas. Rinse with water and drain; then add all vegetables, stock, and 1 cup water only. Cook until vegetables are soft. Tahini is like soft peanut butter, made of sesame, and it is sold in Greek, Italian, or Middle Eastern delicatessens or in the gourmet section of many local supermarkets. In the absence of tahini, use 1 tablespoon of pure sesame oil.

Preparation time: 20 min.   Cooking time: 2 hrs.   Yield: 4 servings

Fish Soup
(*Psarosoupa*)

Aegean Sea Island's delight. This is a very aromatic and very tasty soup, traditionally made out of small fish. However, this recipe is adapted to use fish available in the local market.

    4 ounces codfish, filet
    4 ounces sea bass, filet
    4 ounces scrod, filet
    4 ounces salmon, filet
    2 large carrots, peeled, quartered
    2 medium-size celery stalks, cut in half
    1 medium onion, peeled, cut in half
    1 red bell pepper, sliced, seeds removed
    2 small zucchini, cut in half
    2 medium potatoes, quartered
    5–6 strings parsley
    1/4-cup extra virgin olive oil
    Salt, pepper
    1/3-cup rice
    1 lemon, juice of
    2 tablespoons parsley, finely chopped
    Thin Avgolemono sauce, see page 91

Rinse fish, drain, and set aside. Keep skin on the fish (it will help you to handle fish when cooked).

In a large saucepan, add 6 cups water and bring to a boil.

Add carrots, celery, onion, bell pepper, zucchini, potatoes, parsley strings, half of the oil, salt, pepper, and simmer for 20 minutes or until vegetables are soft.
With a large perforated spoon, remove vegetables and set aside to cool. Discard parsley strings.

Add fish, starting with cod, then scrod, sea bass, and lastly, salmon. Simmer for 5 minutes. Do not overcook, as fish will flake and will be difficult to remove.

Fish Soup (Con't)

With large perforated spoon, remove fish and set aside. Handle fish carefully. Using cheesecloth strain 4–5 cups of broth into another medium saucepan and bring to a boil. Add rice and simmer until soft. Remove from heat.

Use broth from saucepan with rice to prepare Avgolemono sauce. Add to rice and mix. Return to heat and simmer for 2–3 minutes over a low heat until soup becomes thickened.

Remove bones and skin from fish. Arrange fish in the center of a serving platter. Cut vegetables into small pieces and arrange around the fish. Sprinkle with lemon juice, remaining oil, parsley, and pepper. Serve with the soup.

Place soup in individual soup bowls, sprinkle with pepper, and serve hot.

Tip: Because the fish will cook very quickly and then will flake, I suggest that you cook the fish one at a time. This will prevent the fish from flaking and separating and will allow more room in the saucepan to handle this delicate fish.

Preparation time: 40 min.   Cooking time: 35–45 min.   Yield: 4–6 servings

Tripe Soup
(Patsas)

This was my Friday breakfast during my high school years. My uncle, who owned the Scranton restaurant in the town of Plomari, Mytilini, prepared 10 gallons of this soup every Friday with pride, and this is the way he did it for more than 60 years.

   1 pound beef leg
   1 pound beef tripe, frozen
   1 medium-size onion, peeled, cut in half
   1 lemon, quartered
   Salt, pepper
   Pinch of hot paprika (optional)
   6 slices of crusty bread
   Thin Avgolemono sauce, see page 91
   Scordoxido sauce, see page 93

Ask your butcher to cut the beef leg into about 2 x 2-inch pieces.

Wash tripe and leg thoroughly; drain. Cut tripe into 2 x 2-inch pieces.

In a large saucepan, place tripe, leg pieces, onion, half the lemon, and enough water to cover all of the ingredients. Bring to a boil. Reduce heat, cover, and simmer for about 30 minutes. Remove from heat and drain. Discard onion and lemon.

Place tripe on a cutting board and cut into bite-sized pieces. Remove as many bones as possible at this time. Return tripe and leg pieces, remaining lemon, salt, and pepper into same saucepan. Add water to cover all ingredients and bring to a boil. Reduce heat, cover, and simmer until tripe is soft (about 2 hours). Remove from heat and set aside to cool. Remove tripe, leg pieces, and save stock. At this time, remove all remaining bones and discard. Discard lemon pieces also. Pass stock through cheese-cloth to remove any small bones. Return stock to saucepan.

Use hot broth to make Avgolemono sauce, mix with soup, and simmer for about 2–3 minutes. Add tripe and leg pieces and mix. Pour soup into a serving bowl, sprinkle with pepper, paprika if desired, and Scordoxido sauce. Serve hot with bread slices.

Tip: If beef leg is not available then use only the tripe, but you will need 2 pounds.

Preparation time: 25 min.   Cooking time: 2 hrs. 20 min.   Yield: 6 servings

SAUCES

There are few sauces in the Greek cuisine, but the most famous is the Avgolemono sauce; a creamy sauce of lemon juice, eggs, and stock, thickened with cornstarch. This mixture is added to various soups as a thickener or is poured over stuffed grapevine leaves to add flavoring.

Ladolemono (olive oil and lemon) and Ladoxido sauce (olive oil and red wine vinegar) sauces are also essential for various Greek recipes. Ladolemono is light and tasty. It is used to enhance many broiled seafood dishes like fish, shrimp, and lobster. Ladoxido sauce is a very simple, tangy sauce, and one that may be used as a dressing for many different types of salads.

Tzatziki sauce (yogurt flavored with garlic) is served on fried eggplant, rice, or Souvlaki.

Skordalia sauce (ground garlic and pine nuts, mixed in extra virgin olive oil and thickened with bread) is served with fried fish, green bean salad, or beet salad.

Creamy Garlic Sauce
(Skordalia)

Traditional Greek sauce served with boiled red beets, fried eggplant, zucchini, or fried bakala.

> 10 ounces (10 slices) plain, white Italian bread
> 1 medium-size garlic clove, peeled, crushed
> 3 tablespoons pine nuts
> 1 cup extra virgin olive oil
> 2 tablespoons red wine vinegar
> Salt, pepper
> 5 black or green olives, pitted, sliced
> 1 tablespoon parsley, finely chopped

Remove all crust from bread and discard. Toast bread lightly just to dry. Soak bread in water for a few seconds, then squeeze and drain all the water out. Set aside.

In a food processor, beat garlic and pine nuts. Pulse food processor 6–7 times.

Add wet bread and pulse 5–6 times or until mixture is smooth.

Gradually pour oil into mixture and continue beating until entire amount of oil is incorporated and the mixture is smooth.

Add vinegar, salt, pepper, and pulse 2–3 times. Place onto a serving dish and refrigerate for a few hours.

Garnish with olives and parsley before serving.

Tip: You can substitute 2 tablespoons of walnuts for the pine nuts. Walnuts will give a dark color to the Skordalia sauce. Cover and refrigerate leftover Skordalia for one week.

Preparation time: 20 min   Yield: 3 cups

Cucumber and Yogurt Sauce
(Tzatziki)

A very refreshing sauce that goes well with many dishes. My favorite dish is rice pilaf covered with Tzatziki.

  32 ounces plain yogurt
  1 large cucumber, seeds removed, grated and drained
  1 medium garlic clove, crushed and finely chopped
  Salt, pepper
  1 tablespoon red wine vinegar
  1 tablespoon extra virgin olive oil
  1 tablespoon fresh dill, snipped

Strain yogurt through cheesecloth for about 2 hours. Discard water.

Squeeze grated cucumber to remove excess water.

Remove yogurt from cheesecloth and place into a medium mixing bowl. Add cucumber, garlic, salt, pepper, vinegar, and oil. Mix well to combine ingredients. It is a good idea, but not essential, to refrigerate the mixture for a few hours to allow time for the flavors to blend.

Place into a serving bowl and garnish with dill.

Cover and store in refrigerator for 2 weeks.

Tip: Serve with pita bread wedges, fried eggplant, fried zucchini, lamb, or chicken Souvlaki, or just eat it plain.

Preparation time: 15 min.   Yield: 3 cups

Thick Lemon-Egg Sauce
(Avgolemono Sauce)

Eggs combined with broth and lemon make this sauce the most common one used in Greek cuisine. Cornstarch is the main thickening ingredient.

> 2 cups broth
> 1 egg yolk
> 2 1/2 tablespoons cornstarch, dissolved in 1/2-cup cold water
> 1 lemon, juice of
> Salt, pepper

To make thick sauce
In a saucepan, bring broth to a boil and then reduce heat to very low.

In a medium mixing bowl beat egg yolk with cornstarch and lemon juice.

Slowly add about 1 cup of the hot broth while beating the egg mixture.

Pour egg mixture into saucepan and return to medium heat. Stir constantly until sauce thickens. Add salt and pepper; remove from heat, and use as required by the main recipe you are preparing the sauce for.

To make thin sauce
Use 1 tablespoon cornstarch and follow the above instructions.

Tip: Use broth from the recipe for which you are preparing the sauce. If you do not have enough broth, you can supplement with water or chicken broth. You can use water to make the sauce, but it will not be as flavorful. Cornstarch must come to a boil in order for sauce to thicken. The thick sauce is recommended for stuffed grapevine leaves, stuffed zucchini, and stuffed tomatoes, and in any dish where you want the sauce to stay on top of the food. The thin sauce is usually used for soups.

Preparation time: 10 min.   Cooking time: 3 min.   Yield: 2 1/2 cups

Oil and Lemon Sauce
(*Ladolemono*)

This is another fundamental sauce used in the Greek cuisine. It is a simple sauce, but very delicious, especially when it is made with aromatic extra virgin olive oil. It is used for many broiled dishes like fish, shrimp, and lobster.

> 1/4 cup extra virgin olive oil
> 1 lemon, juice of
> Salt, pepper
> 1 tablespoon fresh (parsley or oregano or thyme), coarsely chopped

In a small mixing bowl whisk oil and lemon to emulsify.

Add salt, pepper, and fresh herbs of your choice, and mix.

Cover with plastic and refrigerate until ready to use. Mix well before using. Can keep for a couple of weeks.

Tip: Add 1 teaspoon of hot mustard if sharper flavor is preferred. For seafood dishes, add 1 tablespoon capers.

Preparation time: 10 min.   Yield: 1/2 cup

Oil and Vinegar Sauce
(Ladoxido)

This is a very simple sauce and one that is used as a dressing for various salads.

> 1/2 cup extra virgin olive oil
> 1/4 cup red wine vinegar
> Salt, pepper
> 1 tablespoon fresh oregano, finely chopped

In a mixing bowl whisk oil and vinegar. Add salt, pepper, and oregano. Mix and use as required.

Make Ladoxido sauce in advance and shake well before using.

If you plan to keep it longer than a week, skip fresh oregano and add dry oregano.

Preparation time: 10 min.   Yield: Approx. 1 cup

Garlic and Vinegar Sauce
(Scordoxido)

This is another sauce used as a dressing for red beet salad, string bean salad or tripe soup.

> 1/2-cup red wine vinegar
> 1 garlic clove, crushed
> Salt, pepper

In a mixing bowl, add vinegar, garlic, salt, and pepper. Mix and use as desired. Make Scordoxido sauce in advance and shake well before using.

Pour into a bottle and keep for a few weeks.

Preparation time: 10 min.   Yield: 1/2 cup

Pepper Sauce
(Saltsa Piperias)

This sauce combines the flavors of grilled smoky peppers and the sweetness of grilled onion and garlic. It is excellent served with grilled steaks, lamb chops, or with fried eggplants and cauliflower.

> 6 sweet red bell peppers
> 4 large ripe tomatoes
> 1 medium onion, sliced into 1/2-inch rounds
> 1 hot pepper (optional)
> 4 garlic cloves
> 1/4 cup extra virgin olive oil
> 1 tablespoon parsley, finely chopped
> Salt, pepper

On a hot grill, arrange bell peppers, tomatoes, onion, and hot pepper. Wrap garlic cloves in aluminum foil and place on grill. Remove garlic in 25 minutes and set aside to cool. Remove tomatoes from grill as soon as skin starts to split. Do not overcook.

When peppers and onions are charred, remove from heat. Set onions aside. Put peppers in a paper bag and set aside to cool. Remove and discard skin, stems, and seeds from peppers. Cut peppers into small pieces and set aside. Keep hot pepper separate.

Peel garlic and set aside. Remove and discard skin and seeds from tomatoes; cut into small cubes and set aside.

In a medium saucepan, heat oil; add tomatoes, parsley, salt, and pepper. Simmer until all water evaporates and mixture is left with oil only. Mix frequently.

Add red bell peppers, onion, garlic, as much hot pepper as you wish, and simmer until all water evaporates again. Mix frequently. Remove from heat and set aside to cool. Discard remaining hot pepper.

Pass through a fine food mill and then through a fine sieve to make a smooth sauce. If you prefer a thicker sauce, simmer again over a low heat without a lid to evaporate the excess water. Cover and keep refrigerated for 2 weeks.

Preparation time: 25 min.   Cooking time: 1 hr. 10 min.   Yield: 2–3 cups

Béchamel Sauce

This is the classic béchamel sauce, but made with aromatic extra virgin olive oil instead of butter. You may substitute butter for the oil.

 4 cups (1 quart) milk, hot
 2 tablespoons extra virgin olive oil (or 2 ounces butter)
 4 tablespoons flour
 2 tablespoons cornstarch, dissolved in 3 tablespoons cold milk
 Salt, white pepper
 1/4 teaspoon ground nutmeg (optional)
 1 egg yolk (if called for in menu recipe for which the sauce is being made)

In a medium saucepan, heat oil or melt butter, if using. Add flour and mix constantly for 2–3 minutes or until smooth. Maintain medium heat. You want flour to cook, but not to get dark. Remove from heat and gradually add hot milk, stirring constantly with a wooden spoon. Return to low heat, stirring constantly to prevent lumps.

When smooth and without lumps add cornstarch and mix well. Bring to a boil, stirring all the time. When the sauce thickens, remove from heat, add salt, pepper, nutmeg if desired, and egg yolk. Mix and use as desired.

Tip: Sauce should be smooth and free of lumps. If you have lumps, pass the sauce through a fine sieve. If you want to keep the sauce hot, cover it with plastic wrap and place over hot water.

Preparation time: 10 min.   Cooking time: 20 min.   Yield: 4 cups

Tomato Sauce with Garlic
(Saltsa me Skordo)

This is one of the few tomato sauces used in the Greek cuisine. Although Greeks use tomato as a main ingredient in their recipes, prepared tomato sauce as an ingredient is not very common. This sauce is great for spaghetti or fried eggplants.

    2 tablespoons extra virgin olive oil
    1 small onion, finely chopped
    4 garlic cloves, crushed
    1 carrot, chopped
    1/2-cup celery stalk, chopped
    3 tablespoons parsley, chopped
    1 tablespoon flour
    1/2 (14 ounces) can tomato puree
    1/2-cup white wine
    1 cup water
    1 teaspoon sugar
    2 bay leaves
    Salt, pepper
    3–4 fresh basil leaves, coarsely chopped

In a medium saucepan, heat oil and sauté onion for 2 minutes. Add garlic, carrot, celery, parsley, and sauté for 2 minutes. Add flour, and brown for 1 minute mixing well to coat all vegetables. Add tomato puree, wine, water, sugar, bay leaves, salt and pepper; reduce heat, cover and simmer until vegetables are soft and sauce thickens.

Remove from heat and discard bay leaves. Pass through food mill and through sieve to make smooth tomato sauce. Discard solids left in food mill.

Add basil leaves and mix. If you want thicker sauce return to heat and simmer on low heat without lid to evaporate excess water.

Keep refrigerated for one week.

Preparation time: 15 min.   Cooking time: 1 hr.   Yield: 3 cups

Mayonnaise
(Mayioneza)

There is nothing better than homemade mayonnaise. Aromatic extra virgin olive oil combined with egg yolks, mustard, and lemon makes this the perfect sauce to accompany your boiled fish or boiled chicken.

    2 egg yolks
    1/2 teaspoon sugar
    1/8 teaspoon salt
    1 tablespoon white vinegar
    1 cup extra virgin olive oil
    2 tablespoons lemon juice
    1 tablespoon fresh thyme, finely chopped

Using an electric hand mixer, beat egg yolks, sugar, and salt for 3 minutes at medium speed.

Add vinegar and beat for 2 minutes; gradually add (teaspoon at a time) the oil and keep beating. When oil is added, gradually add lemon juice. Gradually reduce mixer speed to a stop.

Remove from mixing bowl and place into a serving bowl. Add thyme and mix lightly.

Cover and refrigerate. Keep refrigerated for one week.

Preparation time: 10 min.   Yield: 1 1/2 cups

VEGETABLES & RICE

The combination of olive oil and a variety of fresh, seasonal vegetables provides an extremely interesting opportunity for vegetable dishes.

Greeks like to eat many different seasonal vegetables. Leeks, cabbage, and cauliflower, for example, are considered to be winter vegetables. Artichokes, lettuce, and fava beans, are spring vegetables. Summer offers great opportunities for preparing wonderful dishes using tomatoes, zucchini, eggplants, string beans, okra, and other summer vegetables.

Dry goods like beans, lentils, and chickpeas, are served year-round, although many will say that these dishes are for cold weather. This does not mean that you cannot cook a particular vegetable favorite when it is out of season. You can use any vegetable at any time. However, to get the most flavor, aroma, and sweetness from the wide variety of vegetables available, it is best to prepare each vegetable when it is in season.

Greeks are very resourceful when it comes to utilizing their vegetables. They use them to make various vegetarian dishes or combine them with meat and poultry. Eggplant, for example, is very popular, and in this book I have included many recipes that feature eggplant as the main ingredient.

In Greek cuisine, vegetables and olive oil are almost synonymous. They simply go together. You cannot cook Greek vegetable recipes without using good, aromatic olive oil. In fact, some Greek dishes require a good amount of olive oil.

People who like vegetables will love these recipes, and will have the opportunity to explore many new ways of cooking their favorite vegetables using olive oil.

Rice with Fresh Tomato
(*Rizi me Freskia Domata*)

This is an excellent dish to serve during summer, when tomatoes are sweet. This refreshingly light dish can be served as a side dish or as a lunch or dinner entree.

    1 cup long-grain rice
    Salt
    4 tablespoons extra virgin olive oil
    1 tablespoon tomato paste
    2 medium-size ripe tomatoes, seeds removed, grated
    1 cup chicken broth
    Pepper
    2 tablespoons Kefalotyri cheese
    3–4 fresh basil leaves, coarsely chopped

In a medium mixing bowl, add 2 cups hot water, 1 tablespoon salt, and rice; soak for 1 hour. Drain and discard water. Set aside

In a medium saucepan, heat 2 tablespoons of the oil; add tomato paste and sauté for 1 minute, mixing frequently. Add tomatoes and sauté for 2 minutes.

Add chicken broth, 1 cup water, salt, pepper, rice, and remaining oil and bring to a boil. Reduce heat, cover, and simmer until rice is cooked.

Sprinkle with cheese and basil leaves. Serve hot or cold.

Tip: You may substitute 2 cubes of vegetable bouillon dissolved in 1 cup of water for the chicken broth.

Preparation time: 10 min.   Cooking time: 40 min.   Yield: 4 servings

Rice Pilaf
(Pilafi)

Rice pilaf is served as a side dish to accompany many Greek recipes.

    1 cup long-grain rice
    2 (10 ounces) cans chicken broth
    1/4-cup extra virgin olive oil (or 2 ounces butter)
    Salt, pepper

Wash and drain rice with cold water.

In a medium saucepan, bring broth to a boil. Add oil or butter, rice, and salt. Cover with kitchen towel and lid, and simmer on low heat until rice is cooked. Do not mix while cooking. It takes about 25 to 30 minutes to cook. After 25 minutes, remove lid and towel and observe rice mixture. When you see holes and very little steam coming out, this is an indication that rice is cooked.

Sprinkle with pepper and serve hot with meat or chicken dishes. You can also serve with just plain yogurt.

Tip: You may substitute 2 cubes of vegetable bouillon dissolved in 2 cups of water for the chicken broth.

Preparation time: 10 min.   Cooking time: 30 min.   Yield: 4 servings

Rice with Raisins
(Atzem Pilafi)

This recipe incorporates nuts, raisins, and spices. The result is a very flavorful dish, which can be served as a side dish or as a light lunch or dinner. It is also the basic recipe for stuffing tomatoes, eggplants, or bell peppers. It is an excellent side dish for Souvlaki, meatballs, or grilled lamb chops.

1 cup long-grain rice
4 tablespoons extra virgin olive oil
1 small onion, grated
2 tablespoons pistachio nuts, coarsely chopped
2 tablespoons pine nuts
1 teaspoon tomato paste, dissolved in 3 tablespoons water
2 tablespoons black seedless raisins or currants
1 teaspoon ground allspice
Salt, pepper
2 cups chicken broth
2 tablespoons dill, chopped

Wash rice with water; drain and set aside.

In medium saucepan, heat 2 tablespoons of the oil and sauté onions for 2 minutes.

Add pistachio nuts, pine nuts, rice, and brown for 1–2 minutes mixing frequently.

Add tomato paste, raisins, allspice, salt, pepper, and brown for 2 minutes, mixing frequently. Add broth and remaining oil, cover and simmer until rice is cooked.

Mix in fresh dill before serving.

Tip: You may substitute 2 cubes of vegetable bouillon dissolved in 2 cups water for the chicken broth.

Preparation time: 20 min.   Cooking time: 35 min.   Yield: 4–5 servings

Stuffed Tomatoes with Rice
(Domates Gemistes Laderes)

This is an excellent vegetarian dish. The flavors of the extra virgin olive oil combined with those of the baked tomatoes and fresh herbs make this dish irresistible. Serve as an appetizer, as a main dish, or even as a side dish.

　　10 medium tomatoes, ripe but firm
　　5 medium potatoes, peeled and quartered

Stuffing
　　4 medium onions, grated
　　1/2 cup extra virgin oil
　　1/3 cup long grain rice
　　1 tablespoon sugar
　　3 tablespoons parsley, finely chopped
　　2 tablespoons fresh mint, finely chopped
　　3 tablespoons pine nuts
　　3 tablespoons small seedless raisins or currants
　　1/2-cup water
　　Salt, pepper
　　5 tablespoons bread crumbs

Prepare tomatoes by partially slicing the bottom of the tomato (opposite end from stem) without detaching it. Flip over lid and carefully, with a small spoon, scoop out pulp. Preserve tomato juices. Try not to break skin. Discard seeds, chop pulp coarsely, and set aside.

Lightly salt inside of tomatoes and invert into a baking pan to drain for about 30 minutes.

In the meantime, in a small saucepan, bring 3 cups water to a boil and blanch onions for 1 minute. Remove from heat and drain.

In a large frying pan, heat half of the oil and sauté onions. Add rice and brown for 2–3 minutes, mixing frequently. Add tomato pulp, sugar, parsley, mint, pine nuts, raisins, water, salt, and pepper, and give them a good stir. Cover and simmer until rice absorbs all liquid, mixing frequently.

Remove from heat, add 4 tablespoons bread crumbs, and mix. Bread crumbs should absorb all juices. Set aside for 10 minutes to cool.

Preheat oven to 350° F. Rinse tomatoes to remove salt and drain.

Stuff tomatoes lightly (do not overstuff), to allow room for rice to expand. This way the tomatoes will not burst when the rice is cooked. Arrange tomatoes into a 12 x 9 x 2-inch baking pan and place potatoes between tomatoes. Sprinkle tomatoes and potatoes with remaining oil and preserved tomato juices.

Sprinkle tomatoes only with remaining bread crumbs and bake for 90 minutes, basting frequently with juices from the baking pan.

Remove from oven and serve hot.

Tip: Same stuffing may be used to make stuffed peppers, or stuffed eggplants

Preparation time: 40 min.   Cooking time: 1 hr. 30 min.   Yield: 5 servings

String Beans with Onions and Tomatoes (Fasolakia Yahni)

String beans are cooked in olive oil with onions and fresh tomatoes. They are very good as a side dish or as a main course.

    1 pound string beans, fresh or frozen
    1 medium onion, coarsely chopped
    3 medium, ripe tomatoes, cubed
    4 tablespoons tomato puree
    1 teaspoon sugar
    4 garlic cloves, quartered
    1/3-cup extra virgin olive oil
    Salt, pepper
    5–6 strings parsley

If you are using frozen string beans, thaw, wash with cold water, drain, and set aside. If you are using fresh string beans, cut ends, remove strings, and discard. Wash with water and drain.

In a medium saucepan, place half of the string beans, then add half of the onions and half of the fresh tomato and half of the tomato puree. Repeat with the remaining string beans, onions, tomato, and tomato puree. Add sugar, garlic, oil, salt, pepper, and 2 cups water (use 1 cup if you are using frozen string beans). Add parsley strings on top of the string beans.

Cover and simmer over low heat until string beans are soft. This dish is best when the only juice left is the olive oil. If you have water left in the saucepan, remove lid and simmer until juices evaporate.

Remove from heat, discard parsley strings, and serve.

Tip: String beans should be tender. One way to test is to bend them. If tender, they will snap in half.

Preparation time: 20 min.   Cooking time: 1 hr.   Yield: 4 servings

Eggplant with Onions
(Imam Bayildi)

This is a very tasty dish. The combination of eggplant and onions with plenty of extra virgin olive oil, cooked over a low heat, in very little water, makes this dish extremely rich in flavors.

    4 medium eggplants
    Salt
    1 pound onions, thinly sliced
    3 large tomatoes, peeled and chopped
    6-8 garlic cloves, sliced
    1 cup fresh parsley, chopped
    2 teaspoons sugar
    1 cup extra virgin olive oil
    1/4-cup water
    Pepper

Peel a thin layer (skin) of the eggplant lengthwise to form strips. Then insert a knife about 1/2 inch from the bottom of the eggplant and cut through to the other end, stopping 1/2 inch from the end. Do not separate pieces. Turn eggplant and cut through the other sides the same way as you did before. The cuts will allow you to form pockets and stuff the eggplant. Open the slits and sprinkle with salt. Set aside for 30 minutes to drain.

In a medium bowl sprinkle onions with plenty of salt and set aside for 15 minutes.

Squeeze onions, rinse with water, and drain. In the same mixing bowl combine onions with tomatoes, garlic, parsley, sugar, and half of the oil, salt, and pepper.

Rinse eggplant with plenty of water to remove all salt and stuff all openings with the onion mixture. You should be able to incorporate all the vegetables into the eggplant pockets. You need to be careful not to separate the eggplant pieces. If the eggplant pieces separate, the stuffing will not stay in the pockets. If this happens, use toothpicks to hold the pieces together.

In a large saucepan, arrange the eggplants, add the water, and pour over the remaining oil. Cover, bring to a boil, reduce heat to very low, and simmer until all juices are absorbed and eggplant is soft. Remove toothpicks, if you are using, and serve warm or cold, as a main course or as a starter.

Eggplant with Onions (Con't)

Tip: The best eggplants for this dish are the medium or small eggplants found in supermarkets under the name "Italian eggplant." If eggplants are too small, then use more eggplants or cut down on the amount of stuffing.

Preparation time: 30 min.   Cooking time: 2 hrs.   Yield: 4–6 servings

Eggplant with Onions.

Okra with Tomato Sauce
(Okra ston Fourno)

For most Greeks, this is a summer dish that is both popular and versatile. Okra is sold in different sizes, but the preferred size for Greek recipes is the smaller one, which is tastier and more tender. Large okra tend to be tough and chewy.

    2 pounds okra, fresh or frozen
    1/4 cup red wine vinegar
    1/2-cup olive oil
    1 medium onion, chopped
    3 garlic cloves, chopped
    1 (14 ounces) can diced tomatoes
    2 tablespoons parsley, finely chopped
    Salt, pepper

If you are using frozen okra, thaw, wash, and drain. If you are using fresh okra, remove cone-shaped ends from okra without cutting through the pods. Discard ends. Wash, drain, and place into a medium mixing bowl. Add vinegar and mix with okra. Set aside for 20 minutes. After 20 minutes, rinse with water and drain.

In a medium saucepan, heat half of the oil and sauté onions and garlic until soft. Add okra and brown for 2–3 minutes, mixing frequently.

Add diced tomatoes and only 1/2 cup of the tomato juice, parsley, salt, pepper, and bring to a boil. When boiled, remove from heat.

Arrange okra in a 12 x 9 x 2-inch baking pan and cover with juices. Pour remaining oil over okra and bake in the oven at 350° F for about 40–50 minutes or until juices are absorbed and okra is soft. Okra should be left with almost no juice.

Serve hot or cold.

Tip: Instead of baking the okra in the oven, you can cook it in a saucepan. Cover saucepan and simmer until okra is tender, but not too soft; it will take about 20 minutes. After 15 minutes of cooking, remove cover and continue simmering until all the water evaporates and only the okra and oil remain.

Preparation time: 20 min.   Cooking time: 40–50 min.   Yield: 4 servings

Mixed Vegetable Casserole
(Lahanika Tourlou)

You can make this recipe anytime. However, it is best when made with fresh summer vegetables. This is an excellent recipe for vegetarians.

> 1 medium eggplant, cut into large cubes
> Salt
> 2/3 cup extra virgin olive oil
> 2 medium potatoes, sliced thick
> 2 large carrots, peeled and thinly sliced
> 10 string beans, cut in half
> 1 small celery root, peeled and sliced to 1/2-inch rounds
> 1 red bell pepper, diced (discard seeds)
> 4 ounces okra, cone-shaped ends removed and discarded
> 10 white button mushrooms
> 2 zucchini, sliced thick
> 2 medium onions, sliced thick
> 4-5 garlic cloves, peeled and halved
> 3 medium ripe tomatoes, peeled and cut into cubes
> Pepper
> 3 tablespoons parsley, chopped
> 4 tablespoons Kaseri cheese, grated

Salt eggplant cubes and set aside to drain for about 20 minutes. Rinse, drain, and dry with paper towel.

In a large frying pan, heat 2 tablespoons of the oil and sauté potatoes, carrots, and string beans together for about 3-4 minutes. Remove and arrange in a 15 x 10 x 2-inch baking pan.

Add 1 tablespoon more oil to the frying pan; and sauté celery root and bell pepper for 2–3 minutes. Remove and arrange in baking pan.

Add 1 more tablespoon of the oil and sauté okra and mushrooms together for 3-4 minutes. Arrange in baking pan.

Add 1 more tablespoon of the oil in frying pan sauté eggplant and zucchini. Remove and arrange in baking pan.

Add 1 more tablespoon of the oil and sauté onions and garlic and arrange with rest of the vegetables.

In the hot frying pan, add tomato pieces and sauté until soft. Place with the rest of the vegetables. Add salt, pepper, remaining oil, and mix all vegetables together.

Cover with aluminum foil. Make a couple of small holes in the foil for the steam to escape and bake at 375° F for about 50 minutes.

After 50 minutes, remove aluminum foil and bake another 20 minutes, or until juices evaporate and vegetables are left with their oil only.

Remove from oven, sprinkle with parsley and cheese. Serve hot.

Tip: You may substitute Parmesan cheese for the Kaseri cheese.

Preparation time: 45 min.   Cooking time: 1 hr. 10 min.   Yield: 6 servings

Spinach with Rice
(Spanakorizo)

To prepare this dish properly it is best to use fresh spinach. You can also prepare it with frozen spinach, if you do not like fresh spinach.

    1 pound fresh spinach, chopped in half
    1/4 cup extra virgin olive oil
    1 small onion, finely chopped
    4 green scallions, coarsely chopped
    3 tablespoons fresh dill, chopped
    Salt, pepper
    1/2-cup (3 ounces) rice
    1 lemon, cut in wedges

In a medium saucepan, heat oil and sauté onions until soft.

Add scallions and dill, and sauté for 2 minutes. Add spinach, salt, and pepper. Stir, cover, and simmer for about 10 minutes, mixing occasionally. Add rice and 1 cup water; mix to incorporate rice into spinach, cover, and simmer until rice is cooked.

Serve with lemon wedges.

Preparation time: 20 min.   Cooking time: 40 min.   Yield: 4 servings

Leeks with Rice
(*Prasorizo*)

This dish is great in the winter. Leeks are sweet and flavorful. If you like scallions, you will enjoy leeks, especially cooked this way.

2 large leeks
1/2-cup extra virgin olive oil
1 tablespoon tomato paste, dissolved in 3 tablespoons water
Salt, pepper
1/2-cup (3 ounces) rice
4 tablespoons fresh dill, chopped

Cut leeks into 1-inch-long pieces and rinse three to four times in cold water to remove any soil or sand.

In a medium saucepan, bring 3 cups water to a boil and blanch leeks for 1 minute; remove and drain well.

Wipe saucepan dry, heat half of the oil and sauté leeks. Add tomato paste, 1 1/2-cups water, salt, and pepper. Cover and simmer for 10 minutes.

In the meantime, wash rice with water and drain. Add rice and dill to saucepan; mix to incorporate rice with leeks. Cover and simmer over low heat until rice is cooked.

Remove from heat and serve hot.

Tip: Select leeks with long white stems. Use only the tender green part of the leeks. Sometimes you may have to slice leeks lengthwise to ensure better cleaning and removal of sand and soil.

Preparation time: 20 min.   Cooking time: 40 min.   Yield: 4 servings

Artichokes with Lemon-Egg Sauce
(Anginares me Avgolemono)

This dish not only has a delicious flavor, but its creamy sauce is ideal for dipping your crusty bread into. Preparing and cooking artichokes this way, you do not have to deal with hard and chewy leaves as all the hard, chewy parts are removed before cooking. You will love this dish.

        4 large artichokes
        2 lemons
        1/4 cup extra virgin olive oil
        5 shallots, peeled, whole
        2 large carrots, peeled and sliced
        2 small potatoes, peeled, quartered
        8 white button mushrooms, whole
        1 teaspoon flour
        Salt, pepper
        2 tablespoons fresh dill, chopped
        1 cup thin Avgolemono sauce, see page 91

Cut one lemon in half. Pour some salt (about 2 tablespoons) into a small dish and set aside. You will need to rub the artichokes with salt and lemon while cleaning.

To clean the artichokes, start by breaking off the outer 4 or 5 layers of leaves, bending leaves towards the stem, away from you. Cut the tip end of the leaves to expose the center purple choke of the artichoke. Using a small teaspoon, remove the purple choke. Dip lemon in salt and rub exposed center of the artichoke.

With a sharp paring knife, peel off the hard part of the body until most of the green is removed and the white flesh of the artichoke appears. Again, rub with salt and lemon.

Now, cut the top of the stem to about 1 inch long and peel off all hard parts until the white flesh appears. Rub again with lemon and salt, and set aside. Continue with cleaning the rest. Set artichokes aside.

In a large saucepan, heat oil and sauté shallots. Add carrots, potatoes, mushrooms, and sauté for 2 minutes. Add flour and brown for 1 minute, mixing frequently. Add 1 1/2 cups water, salt, pepper, and the juice of one lemon. Bring to a boil.

In the meantime, rinse artichokes with water and arrange in saucepan. Cover and simmer until artichokes are soft. When artichokes are cooked, you should have very little water left over.

Using water, prepare Avgolemono sauce. When artichokes are cooked, pour the sauce over the vegetables and shake the saucepan in a circular motion to coat all ingredients. Arrange on a serving dish, sprinkle with dill and serve hot.

Tip: Select fresh and tender artichokes. Globe should be round and closed at the top. Also, when squeezed, they should be firm. To decorate with carrot rings, keep about 2 inches of the large part of the carrots and boil in water until soft. Slice carrots into 1/4-inch-thick rounds and using a paring knife or an apple corer, make a hole in the center of each carrot slice. Place carrots on the artichoke stem. If artichoke stem breaks, use a toothpick to keep it together, but you must ensure that it is removed before the artichoke is eaten.

Preparation time: 40 min.   Cooking time: 40 min.   Yield: 4 servings

Artichokes with Fava Beans
(Anginares me Koukia)

This is a traditional dish of the island of Mytilini, commonly served in the months of March and April, when artichokes and fava beans are fresh and plentiful. An excellent combination to serve to celebrate the arrival of spring.

    4 large artichokes
    1 lemon, cut in half
    Salt
    1 pound fresh fava beans, deveined
    1/4 cup extra virgin olive oil
    1 medium onion, finely chopped
    1/2 cup scallions, coarsely chopped
    1 teaspoon flour
    3 tablespoons fresh dill, snipped
    1 lemon, juice of
    Pepper
    1 cup thick Avgolemono sauce, see page 91

Cut one lemon in half and in a small dish pour some salt and set aside. You will need to rub artichokes with salt and lemon while cleaning.

To clean the artichokes start by breaking the outer 4 or 5 layers of leaves, bending leaves towards the stem, away from you. Cut the tip end of the leaves to expose the center purple choke of the artichoke. Using a small teaspoon, remove the purple choke. Dip lemon in salt and rub exposed center of the artichoke.

With a sharp paring knife, peel off the hard part of the body until most of the green is removed and the white flesh of the artichoke appears. Again, rub with salt and lemon. Now, cut the top of the stem to about 1 inch long and peel off all hard parts until the white flesh appears. Rub again with lemon and salt. Continue with cleaning the rest. Set artichokes aside.

Wash fava beans, cut in half and set aside.

In a large saucepan, heat oil and sauté onions until soft. Add scallions and sauté an additional 2-3 minutes. Then add flour and brown for 1 minute, mixing frequently. Add 2 cups water, mix well, and bring to a boil.

Wash artichokes with plenty of water, drain, cut in half or quarters, and arrange in saucepan. Arrange fava beans around and between artichokes. Add salt, pepper, and 2 tablespoons of dill. Cover and simmer over low heat until vegetables are soft. Food is done when vegetables are soft and you have about 1 cup of juices remaining in the saucepan. Remove from heat.

Using the juices from the saucepan, make Avgolemono sauce. Pour sauce into artichokes and fava beans and gently mix to coat with sauce.

Sprinkle with remaining dill, add more pepper, and serve hot.

Tip: Fresh fava beans should be tender and bright green. Artichokes should be firm.

Preparation time: 50 min.   Cooking time: 40 min.   Yield: 4 servings

Oven-Roasted Potatoes
(Patates Fournou)

This is a simple, but versatile dish. It is an excellent accompaniment for many dishes in the Greek cuisine. It goes with almost everything.

> 3 pounds potatoes (any variety), peeled and quartered lengthwise
> 1/2 cup extra virgin olive oil
> 2 lemons, juice of
> 3 tablespoons fresh rosemary, finely chopped
> Salt, pepper
> 1 teaspoon paprika

Rinse potatoes with water, drain, and dry with paper towel.

In a large frying pan, heat 4 tablespoons of the oil and brown potatoes; all sides should be coated with oil.

Arrange potatoes in a greased 12 x 9 x 2-inch baking pan. In a small mixing bowl combine remaining oil with lemon juice and mix in rosemary. Coat potatoes with this mixture. Sprinkle with salt, pepper, and paprika.

Bake at 375° F for 60–70 minutes, or until potatoes are golden.

Serve hot with beef, lamb, or pork dishes, or plain.

Tip: You may substitute 2 tablespoons dry oregano and 4 garlic cloves, finely chopped, for the rosemary; add in step 3.

Preparation time: 20 min.   Cooking time: 1 hr. 10 min.   Yield: 6 servings

Mashed Potatoes
(Patates Poure)

I like to make my mashed potatoes using a food mill. This way my potatoes do not have lumps, but they have a coarse texture.

    2 pounds potatoes
    2 ounces butter
    1/2 teaspoon ground cloves
    Salt
    2 cups milk, hot
    4 ounces Kaseri cheese (optional)

In a medium saucepan, add potatoes, cover with water, and bring to a boil. Cover and simmer until potatoes are soft.

Remove from heat and drain. While warm, peel and then puree potatoes using a food mill.

Wipe saucepan clean and heat butter; add potato puree, and mix well. Add cloves, salt, and slowly pour in hot milk, mixing constantly, until mixture is thick, but soft. Pour enough milk to make the potatoes the consistency you prefer. Add cheese if desired, mix, and remove from heat. Serve hot.

Tip: You may substitute Romano cheese for the Kaseri cheese.

Preparation time: 20 min.   Cooking time: 30 min.   Yield: 6 servings

Eggplant with Béchamel
(Melitzanes Papoutsakia)

I believe eggplant stuffed with fresh vegetables is a unique recipe. Even if you do not like eggplants, if you like vegetables and herbs, this is the dish for you.

    3 pounds (about 4) medium eggplants
    3 medium onions, finely chopped
    1/2 red bell pepper, coarsely chopped
    1/4 cup extra virgin olive oil
    4 garlic cloves, finely chopped
    1 small zucchini, grated
    1 medium tomato, diced
    Salt, pepper
    3 tablespoons fresh basil, coarsely chopped
    1 cup fresh parsley, finely chopped
    2 tablespoons fresh mint, finely chopped
    1 cup Kefalotyri cheese, grated
    2 eggs, beaten
    Béchamel sauce, see page 95

Slice eggplants in half, lengthwise. Place into a large saucepan, add enough water to cover eggplants, and bring to a boil. Cover and simmer for 5 minutes or until the inside of the eggplant is getting soft. Do not overcook. Remove, drain, and set aside to cool.

Scoop out the inside of the eggplant, forming a pocket, leaving enough flesh around the skin. This will allow the eggplant to maintain its shape. Eggplants will look like boats.

Dice flesh to make 1/2 cup and set aside. Discard remaining flesh and seeds.

Arrange eggplants into a 15 x 9 x 2-inch baking pan and set aside. Now, prepare stuffing.

Wipe saucepan clean, add 3 cups water, and bring to a boil. Blanch onions and red bell pepper for 1 minute; drain and set aside.

Wipe saucepan dry, heat oil, and sauté onions and red pepper until pepper is soft. Add garlic, diced eggplant, zucchini, tomato, salt, and pepper. Sauté for 5 minutes. Now add basil, parsley, and mint; sauté for 2 more minutes and remove from heat.

Add cheese and eggs and mix well. Stuff eggplants with mixture and set aside.

In the meantime, prepare béchamel sauce. Spread béchamel sauce on each eggplant to cover stuffing.

Preheat oven to 375° F and bake eggplants on the middle rack for 40 minutes or until béchamel is golden.

Remove and serve hot or cold. You can serve this dish as a starter or as a main course.

Tip: Use fresh young eggplants. Young eggplants have shiny flesh and they are soft to the touch. You may also substitute sharp cheddar cheese for the Kefalotyri cheese.

Preparation time: 45 min.   Cooking time: 40 min.   Yield: 4 servings

FISH & SHELLFISH

It almost seems that fish were invented in Mytilini. There are many different types of fish available in the marketplace year-round.

As with vegetables, Greeks buy different fish according to the season. For example, they will buy palamida (bonito) in the spring, codfish in the winter, and sardines in late summer and early fall. You will find tsipoures (porgies or sea bream) in the summer. Barbounia (red mullet) are purchased year-round.

Octopus and squid are also plentiful in the winter and spring. However, they are sought after anytime during the year, especially in the summer. Octopus cooked over charcoal is a fantastic meze for ouzo.

Shellfish, like clams or htenia (small scallops), are seasonal, and shrimp is found most of the year; lobsters are becoming hard to find in Mytilini. You need to be very lucky or you need to know a fisherman to get a lobster.

What is unique on the island is that people will not eat fish unless it is caught on the same day that it is to be prepared. Fish are caught at night and fishermen will raise their nets at dawn and bring the fish to the docks. Fish is purchased right on the docks. Fish that are left unsold are put on ice to be delivered to fish stores around the various towns, and are sold the same day.

My father, for example, will not buy fish that is one day old. It has to be hours old only. But, let me tell you, he is right. Fresh fish is so fresh that you taste the freshness and the aroma of the sea. There is no fish smell, just sea freshness.

Very seldom will you find filet of fish sold in Greek markets or restaurants. Fish is sold and cooked whole, unless it is too large, like codfish or palamida (bonito), which are cut into steaks. Also, most of the fish found in the sea around Greece have bones and scales. However, they are very tasty. That is why Greeks cook their fish very simply. The majority of the time, you will find it grilled, served with Ladolemono sauce, fried in olive oil and served with lemon wedges, or baked with tomato, garlic, parsley, and olive oil.

If you like fish, when you visit the island of Mytilini or any other part of Greece, make sure you buy fresh fish. And remember, ouzo and fish are the perfect partners.

Mussels Rice Pilaf
(Mydia me Rizi)

This is a great dish to make for those who like mussels. The juices from the mussels bring freshness to this dish.

3 dozen large mussels
1/2-cup white wine
1/2-cup extra virgin olive oil
1 medium onion, grated
1 cup rice, washed and drained
1 tablespoon tomato paste, dissolved in 3 tablespoons water
1 tablespoon pine nuts
2 tablespoons currants
1 tablespoon sugar
2 tablespoons parsley, finely chopped
2 tablespoons dill, snipped
Salt, pepper
1 lemon, cut into wedges

Scrape mussels and remove any visible hairy tuft. Wash well with cold water, drain, and set aside. Discard broken ones.

In a large saucepan, heat wine and add mussels. Cover and simmer until mussels open; about 7 minutes. Remove from heat, set aside to cool, and discard unopened ones. Using cotton cheesecloth, strain stock; but reserve 2 cups. Discard solids.

Wipe saucepan clean and heat 2 tablespoons of the oil. Sauté onions until soft. Add rice and brown for 2 minutes, mixing frequently.

Add tomato paste, pine nuts, currants, sugar, parsley, dill, salt, pepper, reserved broth, and remaining oil. Cover and simmer over low heat until rice is cooked.

From reserved mussels, remove shells and discard (keep 4–5 mussels with the shells for decoration). Add mussels to the rice and mix. Cover and simmer for 5 minutes.

Remove from heat and pour rice into a mold. Press rice to compact it into the mold, place a serving dish over the mold opening, and turn over. Slowly pull mold upwards to remove. Garnish with shelled mussels, lemon wedges, and serve.

Preparation time: 30 min.   Cooking time: 40 min.   Yield: 4–6 servings

Mussels with Lemon Sauce
(Mydia me Lemoni)

This was the quickest way for a fisherman to cook mussels at the dock. It is modified a bit and it is great as an appetizer or as a light main course.

3 dozen large mussels
1/4-cup olive oil
2 medium onions, grated
3 garlic cloves, finely chopped
1/4-cup white wine
2 tablespoons parsley, finely chopped
4 bay leaves
Salt, pepper
1 lemon, juice of
1 tablespoon corn starch, dissolved in 3 tablespoons water
2 tablespoons fresh oregano, chopped

Scrape mussels and remove any visible hairy tuft. Wash well with cold water, drain, and set aside. Discard broken ones.

In large saucepan, heat oil and sauté onions until soft. Add garlic and sauté lightly. Add wine, parsley, bay leaves, salt, pepper, and mussels. Cover and simmer until mussels open. When mussels are open, sprinkle with lemon juice, mix, and simmer over low heat for another 5 minutes.

Remove mussels from saucepan and reserve. Discard unopened ones.

Strain stock using cotton cheesecloth. Discard solids. Reserve stock.

In small skillet, mix cornstarch with 1 cup of reserved stock. Mix well to break any lumps. Simmer until sauce thickens. Add more broth as required to make the sauce the consistency of a thin, runny pancake batter.

Place reserved mussels into a serving bowl and pour sauce to cover mussels.

Sprinkle with fresh oregano and serve hot.

Preparation time: 20 min.   Cooking time: 15 min.   Yield: 4 servings

Squid Stuffed with Rice
(*Kalamaria Gemista*)

Preparing squid this way is unique to the island of Mytilini where fresh squid is available almost year-round. Recipes from other islands include prunes and cloves, but the dish becomes too sweet.

> 12 fresh squid, about 5–6 inches long
> 1/2 cup extra virgin olive oil
> 1 small onion, finely chopped
> 1/2 cup rice
> 2–3 tablespoons ouzo (optional)
> 2 tablespoons dill, chopped
> 2 teaspoons ground cumin
> 2 tablespoons pine nuts
> 1 tablespoon tomato paste, dissolved in 3 tablespoons water
> Salt, pepper
> 1 tablespoon paprika
> 12–14 wooden toothpicks

Prepare squid by removing and discarding backbone and ink. Remove and discard head, but keep tentacles. Wash with cold water, drain, and set aside. Chop tentacles into 1/2-inch-long pieces; set aside.

In a medium saucepan, heat half of the oil and sauté onions until soft. Add tentacles and brown for 2-3 minutes, stirring frequently. Add rice, 3/4 cups water, ouzo, dill, cumin, pine nuts, tomato paste, salt, pepper, and mix. Cover and simmer until rice is cooked and juices are absorbed. Set aside to cool.

Preheat oven to 375° F. Use small spoon to lightly stuff squid. Stuff squid to about three-quarters full and then pinch open end with a toothpick. To allow for rice to expand, do not overstuff squid.

Arrange stuffed squid into a 12 x 9 x 2-inch baking pan, add 1/2 cup water, and sprinkle squid with remaining oil and paprika. Cover with aluminum foil and bake for 25 minutes. Remove foil and bake another 15 minutes or until squid is browned.

Remove from oven, discard toothpicks, sprinkle with juices from baking pan, and serve hot or cold.

Preparation time: 20 min.   Cooking time: 40 min.   Yield: 6 servings

Octopus with Wine Sauce
(Xtapodi Krasato)

This dish combines the sweetness of fresh tomatoes and the saltiness of the octopus to make the best tomato sauce for dipping fresh crusty bread.

2–3 pounds octopus
1/4-cup extra virgin olive oil
1 cup red wine
4 large ripe tomatoes
1 tablespoon tomato paste, dissolved in 3 tablespoons water
1 teaspoon sugar
3 bay leaves
5–7 peppercorns
Pepper

Prepare octopus by removing ink and head. Cut into 2-inch-long pieces; wash, drain, and set aside.

In a medium saucepan, heat 2 tablespoons of the oil and sauté octopus for 5-7 minutes, until octopus gets a nice pink color and starts to release its juices. Mix frequently. Add wine and 1 cup of water, cover, and simmer until octopus is soft.

Add tomatoes, tomato paste, sugar, bay leaves, peppercorns, and remaining oil. Cover and simmer until juices are absorbed.

Remove from heat; discard bay leaves, sprinkle with pepper, and serve with rice pilaf or just plain with slices of fresh-baked, crusty bread.

Tip: Select octopus with thick tentacles and larger-size suction cups. Taste food before adding salt; octopus can be salty.

Preparation time: 20 min.   Cooking time: 1 hr. 20 min.   Yield: 4 servings

Octopus with Onions
(Xtapodi Stifado)

This dish is found frequently throughout Greece. The most popular is the one with the tomato sauce, although in some of the islands you will find this dish prepared with white sauce.

2–3 pounds octopus
1/2-cup extra virgin olive oil
1 medium onion, chopped
3 garlic cloves, cut in half
1/4-cup white wine
3 tablespoons red wine vinegar
3 bay leaves
1 (6 ounces) can tomato puree
5-7 peppercorns
1 1/2 pounds small white onions, peeled, whole
1 teaspoon ground cumin
Pepper

Prepare octopus by removing ink and head. Cut into 2-inch-long pieces, wash, drain, and set aside.

If you are using frozen onions, thaw, rinse with water, drain, and set aside.

In a medium saucepan, heat half of the oil and sauté octopus for about 5 minutes, until octopus gets a pink color and starts releasing its juices. Add chopped onion, garlic, and sauté for 2–3 minutes.

Add 1/2 cup water, wine, vinegar, bay leaves, tomato puree, and peppercorns. Cover and simmer for about 40–50 minutes.

Now, add the small onions, cumin, and mix with octopus. Cover and simmer until onions are soft. Remove from heat and discard bay leaves.

Sprinkle with pepper and serve hot with fresh wedges of bread.

Tip: Select octopus with thick tentacles and larger-size suction cups. Taste food before adding salt; octopus can be salty.

Preparation time: 20 min.   Cooking time: 1 hr. 20 min.   Yield: 4 servings

Salted Codfish (Bakala) with Skordalia (Bakaliaro Tiganito me Skordalia)

Bakala is salted codfish found in every Greek and Italian delicatessen. It is very popular in Greek cuisine. This dish is especially popular with those who like retsina (Greek white wine with a pine flavor). This dish may be served as an appetizer or main course.

> 1 pound bakala
> 1/2 cup flour
> 1 teaspoon hot paprika (optional)
> 1/2 can beer
> Pepper
> 1 lemon, cut into wedges

For Frying
> 1 cup olive oil
> 1/2-cup flour
> 1 cup Skordalia sauce, see page 89

Make Skordalia sauce. Cover and refrigerate. In a medium bowl, add enough water to cover bakala and soak for at least 24 hours. Replace water 3–4 times to remove all salt. You can soak in milk also. Remove from water, rinse well with water, and drain. Cut into small cubes about 1 x 1 x 1/2 inches and set aside.

In a medium mixing bowl, combine 1/2 cup flour, paprika, beer, and pepper. Stir well to make a thick batter the consistency of pancake batter.

In a large frying pan, heat half of the oil. Toss bakala pieces in flour and dip in batter. Remove excess batter and fry until golden. Remove and place onto a dish with paper towel to drain any excess oil.

Sprinkle with pepper and serve with Skordalia sauce and lemon wedges.
Taste before adding salt.

Tip: Most of the time, bakala is sold without bones; however, it is a good practice to check and remove small bones before cooking. You may substitute flour with corn meal to toss bakala; just toss bakala in corn meal and then fry for a crispier texture. If you do not want to use beer, you can substitute 1/2 ounce fast-acting yeast mixed into 1 cup water. Add flour and let it stand for 20 minutes to rise before use.

Preparation time: 25 min.   Cooking time: 20 min.   Yield: 30–35 pieces

Salted Codfish (Bakala) with Potatoes (Bakaliaros me Patates)

This is my mother's recipe. I find the combination of tomatoes, potatoes, and bakala sprinkled with dry basil leaves, very interesting. This becomes better when you sprinkle this dish with extra virgin olive oil just before serving.

1 pound bakala
1/2-cup extra virgin olive oil
1 small onion, finely chopped
3 garlic cloves, thinly sliced
1 pound potatoes, sliced 1/2-inch thick
4 tablespoons tomato paste, dissolved in 3 tablespoons water
4–5 peppercorns
4–5 fresh basil leaves, coarsely chopped
Pepper

In a medium bowl, add enough water to cover bakala and soak for at least 24 hours. Replace water 3–4 times to remove all salt. You can soak in milk also. Remove from water, rinse well with water, and drain; cut into 4 equal pieces and set aside.

In a medium saucepan, heat half of the oil and sauté onions and garlic until soft. Add potatoes and brown both sides for 2–3 minutes. Add tomato paste, peppercorns, and 2 cups water; cover and simmer until potatoes are soft but not overcooked.

Place bakala on top of potatoes, cover, and simmer for 5–7 minutes.

Remove from heat, sprinkle with basil and pepper. Taste before adding salt.

Sprinkle with remaining olive oil and serve hot.

Tip: Select thick pieces of bakala. The trick to this dish is not to overcook the bakala. If bakala is overcooked, it becomes tough. I suggest that you check bakala within 5 minutes after placing it atop the potatoes. It is ready when it flakes easily when poked with a fork.

Preparation time: 15 min.   Cooking time: 40 min.   Yield: 4 servings

Salted Codfish (Bakala) Salad
(Bakaliaro Salata)

This is an easy dish to make. It is light and can be served as an appetizer or as a light main dish. I prefer this dish cold.

> 1 pound bakala
> 3 ounces string beans
> 2 medium potatoes, peeled
> 1 small red onion, sliced into thin rings
> 3 tablespoons parsley, coarsely chopped
> 1 red bell pepper, seeds removed, julienned
> 4 tablespoons extra virgin olive oil
> 2–3 tablespoons red wine vinegar
> 2 tablespoons fresh thyme, chopped
> Pepper

In a medium bowl, add enough water to cover bakala and soak for at least 24 hours. Replace water 3–4 times to remove all salt. You can soak in milk also. Remove from water, rinse well with water and drain; cut bakala into about 1-inch-square pieces by about 1/2 inch thick and set aside.

In a medium saucepan, bring to boil 5 cups water. Add string beans and boil until soft. Remove, and set aside to cool. In the same water, cook potatoes until soft. Remove and set aside to cool. Cut into small cubes.

In the same water add bakala and simmer for 5 minutes. Do not cook more because bakala will get tough. Remove from heat and drain.

In a medium mixing bowl, combine bakala, potatoes, onion, parsley, string beans, and bell pepper. Add oil, and toss to incorporate all the ingredients.

Sprinkle with thyme and pepper. Taste before adding salt. Serve warm or cold.

Preparation time: 20 min.   Cooking time: 1 hr.   Yield: 4 servings

Salted Codfish (Bakala) Baked
(Bakaliaro ston Fourno)

Bakala is brushed with garlic, olive oil, covered with herbs and bread crumbs, and then baked. This dish is served with green salad or cabbage salad.

    4 (6-ounce pieces) bakala
    1/2 cup extra virgin olive oil
    5 garlic cloves, crushed
    1/2 cup dry bread crumbs, plain
    1 tablespoon dry oregano
    2 tablespoons parsley, finely chopped
    1 lemon, juice of
    2 tablespoons fresh thyme, chopped
    Pepper

In a medium bowl, add enough water to cover bakala and soak for at least 24 hours. Replace water 3-4 times to remove all salt. You can soak in milk also. Rinse well with water, drain, and set aside.

Preheat oven to 375° F.

Combine 3 tablespoons of the oil with garlic and brush over bakala pieces.

Mix bread crumbs with oregano and parsley; cover bakala.

Arrange bakala in a 12 x 9 x 2-inch baking dish and bake for 20 minutes or until bread crumbs start to get golden. Remove from oven and place in serving dishes.

Combine remaining oil with lemon juice and thyme; pour over bakala. Also, use some of the juices from the baking pan to baste cooked bakala. Sprinkle with pepper and serve hot.

Tip: Select bakala pieces about 3/4 inches thick. Remove skin and any visible bones.

Preparation time: 20 min.   Cooking time: 20 min.   Yield: 4 servings

Codfish with Mayonnaise
(Bakaliaros Vrastos me Mayioneza)

This is a light dish. Fresh codfish is boiled and then served with lemon juice and mayonnaise. It is excellent for light lunch, dinner, or as an appetizer.

    2 pounds fresh codfish steaks
    1 small onion, cut in half
    2 carrots, peeled and cut in half
    1 celery stalk, cut in half
    2 small zucchini, whole
    1 medium potato, quartered
    2–3 strings parsley
    Salt, pepper
    1 lemon, juice of
    2 tablespoons capers
    2 tablespoons parsley, coarsely chopped
    1 cup Mayonnaise, see page 97

Prepare mayonnaise and refrigerate. Rinse fish with water and set aside to drain. Keep skin attached.

In a large saucepan, bring 6 cups of water to a boil. Add onion, carrots, celery, zucchini, potato, parsley strings, salt, pepper; simmer for 10 minutes.

Add fish and cook for 4–5 minutes. Remove fish and set aside to cool. Continue cooking vegetables until soft. Drain vegetables and set aside to cool.

When fish is cool to touch, remove skin and bones and arrange in the middle of a serving platter. Arrange vegetables around the fish.

Sprinkle with salt, pepper, and lemon juice.

Spread mayonnaise over the fish and then sprinkle over with capers and parsley. Serve warm or cold.

Tip: Substitute codfish with scrod, sea bass, or striped bass. Use fish broth to make fish soup. Codfish flakes easily; therefore, do not overcook

Preparation time: 15 min.   Cooking time: 20 min.   Yield: 4 servings

Grilled Sardines
(Sardelles stin Skara)

If you visit the island of Mytilini in August or September, you must visit the town of Scala Kallonis. This town is a fishing port in the gulf of Kalloni where you will find the best sardines in the world. This recipe is the one used by Kalloni fishermen to cook their sardines. This dish is great when eaten hot, just as the sardines come off the grill. Just sprinkle with Ladolemono sauce and enjoy a nice meal.

    1 pound sardines, about 20 pieces
    Salt, pepper
    Dry oregano
    1/2 cup Ladolemono sauce, see page 92

Make Ladolemono sauce and set aside.

Clean sardines, remove scales, rinse with water, and set aside to drain.

Arrange sardines on fish grill and brush with oil; sprinkle with salt, pepper and grill each side for about 5–7 minutes over hot charcoal. Remove and place onto a serving dish.

Arrange sardines on a serving dish; mix Ladolemono sauce with oregano, sprinkle sardines with Ladolemono and serve hot.

Tip: Select fresh and medium-sized sardines. When cleaning the fish, you may want to remove the center bone. This is not necessary, but it makes the sardines easier to eat. To remove the bone, hold the sardine behind the head with one hand, and with your other hand, pull on the head and the bone, working your pointer finger behind the bone. Try it a few times; it is not difficult. This is an excellent way to prepare and serve sardines.

Preparation time: 20 min.   Cooking time: 10–14 min.   Yield: 4 servings

Sardines Plaki in the Oven
(Sardelles Fournou)

If you like sardines, you must try this dish. Garlic and sardines go very well together. Serve as an appetizer or main course.

  1 pound sardines, about 20 pieces
  Salt, pepper
  3 garlic cloves, sliced thinly
  1 large ripe tomato, grated
  3 tablespoons parsley, finely chopped
  1/4 cup extra virgin olive oil
  1 lemon, juice of
  1 lemon, sliced
  1 tablespoon dry oregano

Clean sardines, remove scales, rinse with water, and drain.

Preheat oven to 350° F.

Sprinkle with salt and pepper and arrange in a 12 x 9 x 2-inch baking pan.

In medium bowl combine garlic, tomato, parsley, oil and spread evenly over sardines. Sprinkle with lemon juice, and bake on the middle rack for 25 minutes.

Place onto a serving platter, add lemon wedges, sprinkle with oregano, and serve.

Tip: Select fresh, medium-sized sardines. When cleaning fish, you may wish to remove the center bone. This is not necessary but makes eating easier. To remove the bone, hold the sardine with one hand behind the head, and with the other pull the head and the bone, working your index finger behind the bone. Try a few times; it is not difficult. This is an excellent way to prepare and serve sardines.

Preparation time: 20 min.   Cooking time: 25 min.   Yield: 4 servings

Swordfish Kebabs
(Xifias Souvlaki)

This is a delight of the Greek islands. Fresh swordfish caught daily is full of juices. This is a fantastic substitute for those who do not like fish with bones.

    2 pounds swordfish steak
    4–5 cherry tomatoes, large
    3 tablespoons extra virgin olive oil
    Salt, pepper
    1 tablespoon dry oregano
    1 lemon, cut into wedges
    4–5 12-inch metal skewers
    1/2-cup Ladolemono sauce, see page 92

Make Ladolemono sauce and set aside.

Carefully remove skin and bones from fish and cut into about 1 to 1 1/2-inch cubes.

Arrange fish on skewers alternating between fish and tomato pieces, starting with fish and finishing with fish. If metal skewers are not available, use 12-inch wooden skewers soaked in water for at least 30 minutes. Soaking will prevent the wood from burning.

Brush fish with oil, sprinkle with salt and pepper, and grill over charcoal or on a gas grill. Rotate every 2–3 minutes to grill all sides. Brush with oil after each rotation. Fish is ready when you see some charring on the flesh. Do not overcook, as fish will dry up.

Sprinkle with Ladolemono sauce, pepper, oregano, and serve.

Tip: Purchase fresh swordfish only. Avoid fish that looks dry, as this could be an indication that the fish was frozen. Purchase fish in one thick steak and then cut it into pieces. Serve with boiled dandelion salad or green salad or grilled vegetables.

Preparation time: 20 min.   Cooking time: 12–16 min.   Yield: 4 servings

Fried Sand Shark
(Galeos with Skordalia)

This dish is so popular in the islands that when you go to tavernas it is first-come first-serve basis. Sometimes even the locals have difficulties ordering this fantastic dish. Fabulous meze for ouzo. The people on Mytilini love it.

> 2 pounds sand shark
> Salt, pepper
> 1 lemon, cut into wedges

For Frying
> 1 cup olive oil
> 1/2 cup flour
> 1 cup Skordalia sauce, see page 89

Make Skordalia sauce and set aside.

Rinse fish with water and drain. Sprinkle with salt and pepper and toss in flour.

In a medium frying pan heat oil and fry fish on both sides until golden.

Remove from heat and place on absorbent towel to drain any excess oil.

Sprinkle with more salt and pepper and serve with lemon wedges and Skordalia sauce. Serve hot or cold.

Tip: The most popular sand shark for this dish is the spiny dogfish shark. It is bone-free and the texture is similar to haddock. Substitute sand shark with skate wings. Do not overcook, as fish will get dry. Purchase fish with skin already removed. The ideal size fish is about 2 to 3 inches in diameter, but a larger one will do. If you purchase the fish as one piece, cut it into 1/2-inch-thick round pieces.

Preparation time: 30 min.   Cooking time: 20 min.   Yield: 20 pieces

Baked Skate Wings
(Platia ston Fourno)

This is a very popular dish in Greece. It is also becoming popular in the US. You can find skate in many fish markets and it is not difficult to prepare.

    2 pound (4 pieces) skate wings, skin removed
    Salt, pepper
    2 garlic cloves, crushed
    3 tablespoons lemon juice
    2 tablespoons parsley-finely chopped
    4 tablespoons extra virgin olive oil
    2 tablespoons fresh parsley, chopped

Rinse skate wings with water and drain. Sprinkle with salt and pepper and set aside. Preheat oven to 400° F.

Combine garlic, lemon juice, parsley, and 2 tablespoons of the oil.

Arrange skate wings on aluminum foil and brush with oil mixture. Seal aluminum foil so that steam will not escape.

Place aluminum foil into a baking pan and bake for about 25 minutes.

Remove from aluminum foil, place fish onto a serving dish and baste with some of its juices and remaining oil. Garnish with parsley.

Serve with green salad or dandelion salad.

Tip: Purchase only fresh skate fish. If skate wing has an ammonia-like odor, do not consume. Skate is a member of the ray family. The flesh of the brown, or Raja batis, is the most popular for culinary needs.

Preparation time: 15 min.   Cooking time: 25 min.   Yield: 4 servings

Shrimp with Feta Cheese
(Garides Saganaki)

This is a very popular dish in coastal Greece. This dish combines the sweetness of the Aegean Sea shrimp with the saltiness of the Feta cheese to provide a balanced flavor. Serve hot or cold, as a main meal or as an appetizer.

    1 pound (16–20 count) shrimp, peeled
    1 medium onion, finely chopped
    4 tablespoons extra virgin olive oil
    3 garlic cloves, finely chopped
    1/2-green bell pepper, coarsely
    3 medium ripe tomatoes, diced
    4 tablespoons tomato puree
    1 teaspoon sugar
    Salt, pepper
    2 tablespoons ouzo (optional)
    2 tablespoons fresh parsley, coarsely chopped
    6 ounces Feta cheese, crumbled
    1 medium tomato, cut into thin slices
    3–4 fresh basil leaves, coarsely cut

In a medium saucepan, boil 2 cups of water and blanch onions for 1 minute; drain and set aside.

Wipe saucepan clean and heat oil; sauté shrimp for about 2 minutes. Using perforated spoon remove shrimp and set aside.

Add onions, garlic, green pepper, and sauté until pepper is soft. Maintain low heat so garlic will not burn.

Add diced tomatoes, tomato puree, sugar, salt, and pepper; simmer until juices are absorbed. Then add ouzo and simmer for a couple of minutes. Sauce should be thick. In the meantime, preheat oven to 375° F.

Remove saucepan from heat; add shrimp, parsley, and mix to coat shrimp with sauce. With a spoon, arrange shrimp in a 12 x 12 x 2-inch baking pan and pour remaining sauce over shrimp.

Arrange tomato slices on top of shrimp, and bake for about 15 minutes. Remove from oven and garnish with basil leaves.
Serve with rice pilaf.

Tip: Use less salt, as Feta cheese is salty.

Preparation time: 30 min.   Cooking time: 15 min.   Yield: 4 servings

Shrimp with Rice
(Garides me Pilafi)

This dish uses shrimp stock to cook the rice. Then the shrimp are sautéed in extra virgin olive oil and cooked with rice. It is a very light dish.

    1 pound (20–25) shrimp
    1 cup long grain-rice
    1/4 cup extra virgin olive oil
    1 medium onion, finely chopped
    1 tablespoon tomato paste, dissolved in 3 tablespoons water
    1 teaspoon sugar
    Salt, pepper
    3 tablespoons parsley, finely chopped

Remove shells from shrimp and reserve. Devein shrimp, rinse with water and set aside to drain.

In a medium saucepan, bring 2 cups water to a boil and simmer reserved shells for 10 minutes. Save 1 cup of stock and discard shells and excess stock.

Wipe saucepan clean, heat oil and sauté shrimp for 2 minutes, mixing frequently. Using a perforated spoon, remove shrimp and set aside.

In the same oil, sauté onions lightly. Add tomato paste, sugar, reserved shrimp stock, 1 cup water, salt, pepper, cover and simmer until rice is cooked.

When rice is cooked, add shrimp, parsley and mix with rice. Cover and simmer for another 2 minutes.

Remove from heat and serve hot.

Preparation time: 20 min.   Cooking time: 40 min.   Yield: 4 servings

Lobster Tail with Onions
(Astakos Stifado)

It is becoming very difficult to find fresh lobster on the islands. Therefore, when we find one, we make this dish for an unforgettable dining experience.

2 (14–16 ounces) lobster tails
1/2 cup extra virgin olive oil
2 medium onions, finely chopped
2 garlic cloves, halved
1 carrot, chopped
1/2 cup celery, chopped
2 bay leaves
1 tablespoon flour
1/4-cup white wine
8 ounce can tomato puree
1 cup chicken broth
Salt, pepper
Pinch of hot paprika
2 tablespoons ouzo
4 tablespoons light cream

Slice tails exactly at the joints and set aside. Reserve tail ends for decoration.

In a medium saucepan, heat half of the oil and sauté onions and garlic. Add carrots, celery, and bay leaves; sauté for 2 minutes. Add flour and mix until flour becomes a light golden color. Add wine, tomato puree, chicken broth, salt, pepper, and paprika. Cover and simmer until vegetables are soft.

In the meantime, in a medium frying pan, heat remaining oil and sauté both sides of the lobster pieces (about 2 minutes on each side).

Remove from heat and add ouzo. Using matches (be careful with flames), ignite alcohol. When flames are extinguished, shake the pan in a circular motion to coat lobster pieces with the flavors.

Arrange lobster pieces in saucepan with vegetable mixture and mix lightly to coat the lobster pieces. Cover and simmer for about 10 minutes, or until lobster is soft. Remove lid and set aside for 5 minutes to cool.

Lobster Tail with Onions (Con't)

In the meantime, in a small saucepan, add enough water to cover the reserved lobster tail ends and simmer for 5 minutes. Remove, drain, and set aside.

With a perforated spoon, remove lobster pieces and arrange on a serving dish.

Place a sieve or a fine strainer over a small saucepan and pass the vegetables through to separate the liquid. Mash vegetables with a fork to extract all the juices. Discard solids.

If liquid sauce is very thin, simmer at low heat, without cover, until sauce thickens. Just before removing sauce from heat, add 2 tablespoons of light cream, stir, and remove from heat quickly.

Arrange lobster pieces on a platter, add tail ends to resemble the original tail, and pour hot sauce over the lobster tails.

Garnish with remaining cream. Serve with rice pilaf or plain spaghetti.

Preparation time: 30 min.   Cooking time: 40 min.   Yield: 4 servings

Lobster Tail Salad
(Astakos Vrastos Salata)

In most restaurants, boiled lobster is served with hot butter. Greeks like their lobster with Ladolemono sauce. The Ladolemono sauce adds additional flavor and it is an excellent substitute for hot butter.

    2 pounds (2–3) lobster tails
    1 celery stalk, chopped
    1 small onion, cut in half
    2 eggs, hard-boiled, quartered
    2 tablespoons capers
    Salt, pepper
    2 tablespoons fresh parsley, finely chopped
    1/2 cup Ladolemono sauce, see page 92

Prepare Ladolemono sauce and set aside.

Wash lobster tails and set aside to drain.

In a large saucepan, bring 6 cups water to a boil. Add celery, onion, and lobster tails; simmer for 8–10 minutes.

Remove from heat and place lobster into ice-cold water. Discard vegetables.

When lobster is cold, use kitchen scissors to cut and remove the shell. Discard shell, but keep tail end. On a cutting board, slice lobster into 1/2-inch-thick pieces.

Arrange lobster pieces on a serving platter and place eggs around the pieces. Decorate with tail ends. Pour Ladolemono sauce over the lobster, sprinkle with capers, salt, pepper, and parsley. Serve cold.

Preparation time: 20 min.   Cooking time: 10 min.   Yield: 4 servings

Grilled Mackerel
(Skoubria Psita)

Mackerel is an oily fish with no scales and very few bones. It is excellent for grilling. Late spring is the season to enjoy this fish.

    2 pounds (4–5) mackerel
    1 medium onion, sliced thinly
    1 small tomato, diced
    4 tablespoons parsley, coarsely chopped
    2 garlic cloves, finely chopped
    3 tablespoons extra virgin olive oil
    Salt, pepper
    5–6 toothpicks
    3 tablespoons virgin olive oil
    1/2 cup Ladolemono sauce, see page 92

Prepare Ladolemono sauce and refrigerate.

Rinse and drain fish. Sprinkle with salt and set aside. Remove heads and tails, if desired.

In a small mixing bowl, combine onions, tomato, parsley, and garlic; stuff the belly of the fish with the vegetable mixture. Use toothpicks, or staple the opening to prevent the mixture from falling out. Refrigerate for 2 hours to allow the fish to absorb the flavors from the vegetables.

Grill on hot gas or charcoal grill. Brush with olive oil as you grill the fish. Since mackerel is a fatty fish, it has a tendency to flame while grilling, so you will need to watch it closely to keep it from burning.

Remove from charcoal and place onto a serving dish. Remove and discard toothpicks. Serve fish with Ladolemono sauce and green salad or boiled green beans with Skordalia.

Tip: If mackerel is large, make a slit lengthwise on both sides of the mackerel. This will help the grilling process and will allow the fish to cook evenly. Substitute mackerel with large, red mullets. Red mullets have a tendency to flake, so you need to handle them carefully. You do not need to slit red mullets.

Preparation time: 20 min.   Cooking time: 15–20 min.   Yield: 4–5 servings

Red Snapper with Garlic
(Sinagrida Plaki)

This recipe is for those who like fish, but are tired of eating fried or broiled fish. It is an excellent way to prepare red snapper, sea bream, porgies, or even striped bass. The fish is wrapped with onions, tomatoes, and garlic, and then baked for 25 minutes.

2 pounds (4) red snapper
1/4-cup extra virgin olive oil
1 small onion, finely chopped
4 garlic cloves, sliced
1 can (8 ounces) tomatoes, diced
1/2 teaspoon sugar
Salt, pepper
1/2 cup parsley, coarsely chopped
1/4 cup white wine
1 lemon

Rinse fish with water and drain. Sprinkle with salt and set aside. Keep fish whole; do not filet.

In a skillet, heat half of the olive oil and sauté onions and garlic until soft. Add tomatoes, sugar, salt, pepper, parsley, and wine; simmer over low heat without a cover until onions are soft and the liquid is absorbed.

Preheat oven to 350° F.

Place fish into a baking pan and cover with tomato mixture. Slice half of the lemon into thin slices and arrange over fish.

Sprinkle with remaining oil and bake on the middle oven rack for about 25 minutes. Remove from oven and serve hot with boiled dandelion or boiled cauliflower. Cut remaining lemon in half and serve with the fish.

Preparation time: 20 min.   Cooking time: 30 min.   Yield: 4 servings

Porgies with Sun-Dried Tomatoes and Olives (Tsipoures me Elies)

Both porgies and olives are plentiful on the island of Mytilini. Sun-dried tomatoes and garlic, mixed with parsley and extra virgin olive oil are the dominant flavors in this dish.

    2 pounds (3–4) porgies
    1 can (8 ounces) plum tomatoes
    1/4-cup extra virgin olive oil
    1 medium onion, finely chopped
    4 garlic cloves, sliced
    6–8 sun-dried tomatoes, coarsely chopped
    4 tablespoons parsley, finely chopped
    Salt, pepper
    15 black olives, pitted and sliced in half
    1 lemon
    2 tablespoons capers
    2 tablespoons fresh oregano, coarsely chopped

Keep fish whole; do not filet. Rinse with water and drain. Sprinkle fish with salt and set aside. Remove plum tomatoes from can, cut into 4–5 pieces, and discard juices.

In a frying pan, heat half of the olive oil and sauté onion and garlic. Add sun-dried tomatoes, plum tomatoes, parsley, salt, and pepper; simmer over a low heat, uncovered, until juices are absorbed.

Preheat oven to 350° F.

Arrange fish in a baking pan, cover with tomato mixture and olives.

Cut half of the lemon into slices and the other half into wedges. Arrange lemon slices over the fish. Bake on the middle oven rack for 25 minutes.

Remove from oven; add capers and sprinkle with fresh oregano. Serve with lemon wedges and dandelion salad or cauliflower salad.

Tip: Substitute whole porgies with flounder filets or tilapia filets. When using fish filets, sprinkle dry bread crumbs on top of the tomato sauce and bake until bread crumbs are light golden.

Preparation time: 25 min.   Cooking time: 25 min.   Yield: 4 servings

Fried Red Mullet
(Barbounia Tiganita)

Red mullet, especially one called Kotsomoura, is very popular on Mytilini. The fish is small, less than 3 inches long, and it is fried to a crispy texture. When fried in olive oil, the fish is sweet and you can taste the freshness of the sea.

>   1 pound red mullet
>   Salt, pepper

For Frying
>   1/2 cup virgin olive oil
>   1/4 cup flour
>   1 lemon, cut into wedges

Remove scales, clean fish, rinse with water and drain. Sprinkle with salt and pepper; set aside. In a plastic bag, add flour and toss the fish to coat.

In a frying pan, heat oil and fry fish. Fry each side well before turning over; this way the fish will be crispy. Otherwise, it will become soggy. Fish should be golden brown.

Remove from heat and place on paper towel to drain any excess oil. Sprinkle with more salt and pepper.

Serve hot with lemon wedges and tomato salad or dandelion salad.

Preparation time: 20 min.   Cooking time: 25 min.   Yield: 4 servings

Grilled Red Snapper
(Sinagrida Psiti)

Grilling is a very popular method of cooking fish in Greece especially porgies or red snapper. When the fish is fresh you can almost taste the Aegean in every bite.

    2 pounds (4) red snapper
    Salt, pepper
    2 tablespoons extra virgin olive oil
    1/2 cup Ladolemono sauce, see page 92

Make Ladolemono sauce and set aside.

Rinse fish with water, pat dry and sprinkle with salt.

Place fish on a hot, well-oiled, fish rack and grill for 8–10 minutes on each side.

Remove fish from heat and arrange on a serving plate.

Sprinkle with Ladolemono sauce, pepper, and serve with tomato salad.

Tip: Substitute porgies or red mullet for the red snapper. Select larger-sized red mullet. If you are using red mullet, you need to use extra care when removing it from the grilling rack as red mullet flakes easily. Use a nonstick fish rack and try to cook fish well on each side before turning.

Preparation time: 10 min.   Cooking time: 20 min.   Yield: 4 servings

Sepia (Cuttlefish) with Wine
(*Soupies Krasates*)

Sepia dishes are very popular with Greek islanders. The most popular dishes are made with rice, wine, or spinach. Sometimes you will find sepia grilled or fried. In this dish, sepia is slowly cooked, sautéed in wine and allspice.

        3 pounds sepia
        1/4-cup extra virgin olive oil
        2 medium onions, coarsely chopped
        1 tablespoon allspice, whole
        4–5 peppercorns
        8 ounces tomato puree
        2 cups red wine
        1/2-teaspoon sugar
        2–3 bay leaves
        Salt, pepper

To clean sepia, remove soft bone and discard. Discard ink sac and head. Keep tentacles. Wash sepia and tentacles; drain.

Cut sepia's main body into 2-inch squares and the tentacles into 2-inch-long pieces.

In a medium saucepan, heat oil and sauté sepia for 5 minutes. Add onions and sauté for 2 minutes.

Place allspice and peppercorns into a spice bag and add to saucepan.

Add tomato puree, wine, sugar, bay leaves, salt, and pepper. Cover and simmer over low heat until sepia is soft and juices are absorbed. If sepia is cooked but juices are not completely evaporated, simmer a few more minutes uncovered.

Remove from heat and discard bay leaves. Serve hot with rice.

Preparation time: 15 min.   Cooking time: 1 hr. 20 min.   Yield: 4 servings

Sepia with Spinach
(Soupies me Spanaki)

Spinach and sepia go very well together. Sepia is sautéed and then cooked in wine with onions and spinach.

> 3 pounds sepia
> 2 pounds fresh spinach
> 1/2 cup extra virgin olive oil
> 2 medium onions, chopped
> 1 bunch scallions, chopped
> 1/2 cup white wine
> 1/2 cup parsley, finely chopped
> Salt, pepper
> 3 tablespoons fresh dill, snipped
> 1 lemon, juice of

To clean sepia, remove soft bone and discard. Discard ink sac and head. Keep tentacles. Wash sepia and tentacles; drain.

Cut main body into 2-inch squares and the tentacles into 2-inch-long pieces.

Wash spinach with plenty of cold water, cut in half, and let it drain well.

In a large saucepan, heat oil and sauté sepia for 5 minutes. Add onions and sauté for 5 minutes. Add scallions, spinach, wine, parsley, salt, pepper, 1/2 cup water, and mix. Cover and simmer until sepia is soft.

Uncover and cook an additional 10–15 minutes or until juices are absorbed. This dish is done when sepia is soft and olive oil is the only liquid remaining in the saucepan. Remove from heat and place into a serving bowl.

Sprinkle with dill, add lemon juice, and scoop some of the oil from the saucepan over the dish. Serve hot.

.

Tip: Substitute fresh spinach with frozen leaf spinach. If you are using frozen spinach, then cook sepia until it is soft before you add the spinach.

Preparation time: 25 min.   Cooking time: 1 hr. 20 min.   Yield: 4 servings

Sepia (Cuttlefish) with Rice
(Soupies me Mavro Pilafi)

This is a unique dish and it is for those who like bold flavor.

  3 pounds sepia
  1/2 cup extra virgin olive oil
  1 small onion, grated
  1 tablespoon tomato paste, dissolved in 3 tablespoons water
  1/2 cup white wine
  Salt, pepper
  1/2 cup white rice
  2 tablespoons parsley, finely chopped

To clean sepia, remove soft bone and discard. Remove ink sac and reserve.

Remove and discard head, but keep tentacles. Wash sepia and cut main body into 2-inch-squares and tentacles into 2-inch long pieces. Set aside to drain.

In a medium saucepan, heat oil and sauté sepia for 5 minutes. Add onion and sauté for about 2 minutes.

Add sepia ink, tomato paste, wine, 1 cup water, salt, and pepper. Cover and simmer for 30 minutes.

Add rice and mix. Cover and simmer for another 30 minutes or until rice is cooked and all juices are absorbed.

Add parsley; mix and remove from heat. Serve hot.

Tip: Oil will not combine with sepia ink; therefore, ensure that water evaporates before you remove from heat.

Preparation time: 20 min.　　Cooking time: 1 hr. 20 min.　　Yield: 6 servings

Fried Flounder Filet
(Glosses Tiganites)

Flounder filet is dipped in milk, then tossed in corn meal, and fried. Flounder is not as plentiful as other fish on the Greek islands, but when available, this is a very popular recipe.

>   2 pounds (5-7) flounder filets
>   1/4-cup milk
>   Salt, pepper
>   1 lemon, cut into wedges

For Frying
>   1/2 cup olive oil
>   4 tablespoons flour
>   4 tablespoons corn meal (optional)

Rinse flounder filets with water and drain.

In a medium mixing bowl, soak filets in milk for 10 minutes.

Remove filets from milk one at a time, sprinkle with salt and pepper, and toss in flour or corn meal. Discard milk.

In a frying pan, heat oil and fry flounder on both sides until golden. Remove from heat and place on paper towel to drain any excess oil.

Place on a serving platter and sprinkle with salt and pepper. Arrange lemon wedges around fish. Serve hot with lettuce salad or red beet salad.

Preparation time: 20 min.   Cooking time: 15 min.   Yield: 4 servings

A variety of fish available at a fish market in the
town of Plomari, Mytilini.

POULTRY & GAME

Chicken is very popular in Greece. You will find it roasted, stuffed, baked with potatoes, with tomato sauce, and with other vegetables.

Turkey is a bird that is not very popular on the islands; however, in some towns, stuffed turkey is the traditional dish for Christmas Day.

Greeks love to hunt. So game birds like partridge and woodcock are very popular. Both are fall and winter game birds. These recipes are fantastic for those who like the flavor of game birds.

The most popular recipe on the island of Mytilini is free-range rooster cooked with fresh tomato and served with rice. Unfortunately, it is very difficult to find free-range chicken but when you visit the island, ask the locals which taverna offers this dish. Call the taverna in advance and ask them to prepare this dish for you. You will need a big appetite. These birds are big, tender, and very tasty.

Most of the chickens sold on the island are farm-raised and fed proteins to promote fast growth, so taste and flavor are sacrificed.

Stuffed Turkey
(Galopoula Gemisti)

Turkey dishes, in general, are not very popular on the Aegean Islands. However, at Christmastime, some of the villages on the island of Mytilini will prepare this dish. On Christmas Day, the local baker will cook all the turkeys for the whole town.

1 (10–12 pound) fresh turkey
Salt, pepper
3 tablespoons fresh sage, finely chopped
3 tablespoons extra virgin olive oil
1 small onion, cut in half
1 small celery stalk, chopped
1 carrot, peeled, cut in half
Plenty of aluminum foil
Syringe with needle

Rice Stuffing
Gizzards from turkey
1/4 cup extra virgin olive oil
1 medium onion, grated
5-7 chestnuts, baked, peeled and chopped
2 tablespoons pine nuts
1 small, hard apple, peeled and cubed
1/4 cup raisins
3 tablespoons dill, chopped
1 cup long-grain rice
1 cup turkey stock
1 cup water
Salt, pepper

Greek Sausage with Sage Stuffing (optional)
1 pound Greek sausage
2-3 tablespoons extra virgin olive oil
1 medium onion, finely chopped
1/4 cup milk
6 ounces dry bread crumbs
1/4-cup fresh sage, finely chopped
Salt, pepper
1 cup turkey stock

Stuffed Turkey (con't)

Gravy
> 2 cups turkey broth, hot (see this recipe)
> 1 tablespoon flour
> 1/2 cup white wine
> Salt, pepper

If you are using frozen turkey, cover with kitchen towels and thaw.

Wash turkey thoroughly inside and out and set aside to drain and dry. Reserve neck and gizzards.

Mix 1/2 teaspoon salt, 1/2 teaspoon pepper, 3 tablespoons sage, and 1 tablespoon of the olive oil. Very carefully, lift the skin of the turkey from the breast side and spread the sage mixture evenly between the flesh and skin. Pay attention so as not to break the skin.

Place the remaining olive oil into a syringe and inject the turkey meat in various areas. Refrigerate turkey.

In a medium saucepan, bring 6 cups water to a boil. Add neck, onion, celery, carrot, salt, and pepper; cover and simmer until neck meat is soft. Remove from heat and set aside to cool. Remove meat from the neck and reserve. Discard bones. Strain stock and reserve 2 cups. Discard vegetables.

To Make Rice Stuffing

In a small saucepan, bring 2 cups water to a boil. Add gizzards and simmer for about 2 minutes. Remove, drain, and discard stock.

On a cutting board, chop gizzards into small cubes.

In a medium saucepan, heat oil and sauté onion.

Add meat from neck, gizzard cubes, chestnuts, pine nuts, apple, raisins, and dill; sauté for 2 minutes. Add rice and sauté an additional 2–3 minutes, mixing frequently. Add salt and pepper.

Add 1 cup of reserved stock and 1 cup water; cover and simmer until rice absorbs juices. Remove from heat and set aside to cool.

To Make Greek Sausage with Sage Stuffing (if using)

In a medium saucepan, heat oil and sauté onion until soft, mixing frequently. Remove sausage from casing, dice, and brown with onions, mixing frequently.

Add milk, bread crumbs, sage, salt, pepper, and 1/2 cup of the broth. If stuffing is too dry, add more broth.

Remove from heat and let cool.

To Stuff and Cook the Turkey

With large spoon, stuff main body cavity of turkey with rice stuffing. If you have left-over rice stuffing, return to the stove, add 1/4 cup water, if needed cover; and simmer until rice is cooked. Remove from heat and set aside.

Using a needle and string, sew cavity opening closed. Now stuff the neck cavity with Greek sausage stuffing. Using a needle and string, sew cavity opening closed.

Sprinkle turkey with salt and pepper. Preheat oven to 400° F. Place turkey on top of aluminum foil and seal. The foil should be closed tightly, but should not touch the turkey skin on top. Place into a large baking pan.

Bake for 50 minutes at 400° F. Reduce heat to 350° F and cook an additional 3 hours for a 10-pound turkey. Add 15 minutes for each additional pound.

Uncover turkey and cook an additional 50-60 minutes or until turkey is golden brown and the temperature between the body and thigh is 175° F.

Remove from the oven and cool for 20-30 minutes before carving. Strain and save 2 cups of the broth to make the gravy. See directions below.

To Make the Gravy

Skim and discard visible fat from the broth.
In a medium saucepan, brown flour until golden, mixing frequently.
Slowly add hot broth and using a whisk, stir vigorously to break up any lumps.
Add wine, salt, and pepper. Simmer to get the consistency you want.
Serve hot with turkey.

Stuffed Turkey (con't)

Tip: Stuff turkey and cook immediately. If you cannot find Greek sausage, you can make your own. You do not need casing. Just make sausage patties. Here is the recipe:

To Make Your Own Greek Sausage
>   2 pounds ground pork (include some fat)
>   1 pound ground lamb
>   1/2-cup red wine
>   3 tablespoons dry oregano
>   1 tablespoon ground anise seed
>   2 tablespoons orange zest
>   2 garlic cloves, crushed
>   Salt, pepper
>   Sausage casing (optional)

Mix all ingredients together. Shape into patties and refrigerate for 12 hours to blend the flavors. If you have casing then stuff the casing, and keep refrigerated for about 1 week.

Tip: When you lift the turkey out of the pan, tilt it to drain excess juices. It is a good idea to leave the turkey covered for 20-30 minutes to "relax" before carving.

Preparation time: 70 min.   Cooking time: 4–5 hrs.   Yield: 10–12 servings

Turkey with Tomato Sauce
(Galopula Kokinisti)

This is a simple recipe. Turkey breast is sautéed in extra virgin olive oil and then cooked in tomato sauce over a low heat.

 2 pounds turkey breast, sliced into 4 pieces
 Salt, pepper
 2 tablespoons flour
 3 tablespoons extra virgin olive oil
 1 medium onion, grated
 8 ounces tomato sauce
 1/2 cup red or white wine
 2 tablespoons sage leaves, finely chopped

Wash turkey pieces thoroughly and drain. Sprinkle with salt, pepper, and toss in flour. Set aside.

In a medium saucepan, heat oil and sauté turkey on both sides until golden. Remove turkey pieces from the oil. In the same oil, sauté onions until soft.

Add tomato sauce, wine, sage, salt, pepper, and turkey pieces; cover and simmer until turkey is tender and juices are absorbed. Add a little water, if required. Ensure that tomato sauce is thick before serving.

Serve with potato croquettes or rice pilaf.

Preparation time: 20 min.   Cooking time: 1 hr.   Yield: 4 servings

Chicken Kebab
(Kotopoulo Souvlaki)

This dish has become very popular with people watching their diets. Marinate the chicken and grill on charcoal for best flavor.

   1 1/2 pounds chicken breasts
   1/4 cup extra-virgin olive oil
   1/4 cup white wine
   1 small onion, coarsely chopped
   2 tablespoons fresh rosemary, chopped
   8–10 cherry tomatoes
   1 red bell pepper, cut into 1-inch squares
   1 medium onion, quartered, layers separated
   1 lemon, juice of
   Salt, pepper
   8–12 skewers, 12 inches long
   Tzatziki sauce (optional), see page 90

Rinse chicken with water and drain. Cut chicken into approximately 1 1/2-inch cubes and place into a medium mixing bowl.

Make marinade by combining oil, wine, chopped onion, and rosemary. Pour over chicken and mix well to coat the chicken pieces. Cover the bowl with plastic wrap or foil and refrigerate for 3-4 hours. For best flavor, refrigerate at least 12 hours.

Skewer chicken pieces, starting with chicken and alternating with cherry tomato, green pepper, and onion layer. Complete skewer with chicken pieces. This will prevent the vegetables from falling off.

Grill chicken over charcoal, brushing frequently with marinade. Turn to grill all sides. In absence of a gas or charcoal grill, broil in oven, but turn frequently to achieve even cooking.

Remove from heat and sprinkle with lemon juice, salt, and pepper. Arrange on a serving dish; remove and discard skewers.

Serve with Tzatziki sauce and rice pilaf. You can also roll in pita bread with Tzatziki sauce.

Tip: Soak skewers in cold water for 1 hour before using. This will prevent the wooden skewers from burning. You may use metallic skewers, if available.

Preparation time: 25 min.   Cooking time: 20–25 min.   Yield: 4–5 kebabs

Stuffed Chicken
(Kotopoulo Gemisto)

This is a very popular way of cooking chicken in Greece. Fresh rosemary combined with bread crumbs gives a nice aroma to the chicken.

> 1 (3-4 pounds) whole chicken
> 1/4 cup virgin olive oil
> 1 small onion, grated
> 1/2-cup dry bread crumbs
> 1/4 cup white wine
> 2 tablespoons tomato puree
> 1 tablespoon fresh rosemary, finely chopped
> Salt, pepper
> Needle and cotton string
> 2 lemons, juice of
> 2 garlic cloves, crushed
> 1 teaspoon paprika
> 3 pounds potatoes, peeled and quartered

Rinse chicken thoroughly with water, inside and out, and drain. Remove visible fat and discard. Ensure that skin is intact. Dice gizzards and set aside.

In a medium saucepan, heat 2 tablespoons of the olive oil and sauté onions. Add gizzards and sauté for 2 minutes. Add bread crumbs and brown for 1 minute, mixing frequently.

Add wine, tomato puree, rosemary, salt, and pepper; mix well. Simmer for 1 minute; remove from heat. Set aside for 5 minutes to cool.

Preheat oven to 375° F.

With a large spoon, stuff chicken cavity. Use needle and cotton string to close cavity opening.

Arrange chicken in a 12 x 9 x 2-inch baking pan with the breast side up.

In a large bowl, combine lemon juice, remaining oil, garlic, and paprika. Add potatoes and mix well with oil.

Arrange potatoes around chicken and pour remaining oil mixture over potatoes.

Bake for 90 minutes or until chicken is tender. Serve with dandelion salad or green salad.

Preparation time: 20 min.   Cooking time: 1 hr. 30 min.   Yield: 4–5 servings

Chicken with Okra
(Kotopoulo me Bamies)

Okra and chicken go very well together. I prefer using thighs and legs, as they are tender and add more flavor to this dish.

    2 pound chicken (thighs or legs)
    1 pound okra, fresh or frozen
    1/4 cup extra virgin olive oil
    Salt, pepper
    1 medium onion, chopped
    1 (14 ounces) can diced tomatoes
    2 tablespoons parsley, chopped

If you are using frozen okra, thaw, rinse, and set aside to drain. If you are using fresh okra, peel the cone-shaped end, without piercing through. Rinse, drain, and set aside. Wash and drain chicken pieces.

In a large saucepan, heat half the oil. Sprinkle chicken pieces with salt and pepper and brown on both sides.

Remove chicken from saucepan and in the same oil, sauté onions for 1 minute.

Return chicken pieces to saucepan, add tomatoes and their juices, salt, pepper, and remaining oil; cover and simmer until chicken is soft, but not completely cooked (about 50–60 minutes). If you are using frozen okra, then cook chicken well before adding okra. Frozen okra requires about 15 minutes of cooking. Fresh okra requires about 30 minutes of cooking.

Arrange okra between the chicken pieces, cover, and simmer until okra is tender, but firm. Do not overcook okra. Ensure that tomato sauce is thick.

Place chicken in the center of a serving dish. Arrange okra around the chicken. Garnish with tomato sauce. Skim some oil from the saucepan and pour it over the okra and chicken.

Sprinkle with parsley and serve hot with rice pilaf.

Preparation time: 25 min.   Cooking time: 1 hr. 30 min.   Yield: 4 servings

Chicken with Tomato and Olives
(Kotopoulo me Elies)

Chicken is sautéed in virgin olive oil and then cooked in tomatoes, wine, and olives. Delicious!

    4 chicken legs, whole
    3 tablespoons virgin olive oil
    Salt, pepper
    1 tablespoon flour
    1 small onion, finely chopped
    1/4 cup white wine
    1 (14 ounces) can diced tomatoes
    2 tablespoons parsley, coarsely chopped
    10 Kalamata olives, pitted, sliced

Rinse chicken thoroughly with water and set aside to drain.

In a large saucepan, heat oil. Sprinkle chicken legs with salt and pepper; toss in flour and brown on both sides. Remove chicken and set aside.

In the same oil, add and sauté onions until soft, mixing frequently. Add wine, tomatoes, parsley, olives, salt, pepper, and 1/2 cup water. Cover and simmer until chicken is tender.

Remove from heat and serve with rice pilaf.

Preparation time: 20 min.   Cooking time: 1 hr. 30 min.   Yield: 4 servings

Chicken with Onions
(Kotopoulo me Kremmydia)

This is a very tasty dish. The caramelized onions give this dish a subtle sweetness.

    2 pounds chicken, legs and thighs
    1/4 cup extra virgin olive oil
    3 medium onions, sliced
    3 garlic cloves, sliced
    1 tablespoon flour
    1/2 cup red wine
    3 tablespoons vinegar
    8 ounces tomato puree
    2 bay leaves
    Salt, pepper

Rinse chicken thoroughly with water and set aside to drain.

In large saucepan, heat oil and brown chicken. Brown on all sides. Remove chicken and set aside.

In the same oil, sauté onions until soft, but not browned, mixing frequently. Add garlic and sauté for another 2 minutes.

Add flour and brown until light golden, mixing continually. Add wine and vinegar and bring to a boil; reduce heat and simmer for 2 minutes.

Add tomato puree, bay leaves, chicken legs, salt, and pepper; cover and simmer until chicken is soft.

Remove from heat and serve hot with rice pilaf.

Preparation time: 25 min.   Cooking time: 1 hr. 30 min.   Yield: 4 servings

Grilled Chicken
(*Kotopoulo sta Karvouna*)

The secret for a tasty, tender, juicy chicken is to marinate overnight and then grill over hot charcoal. The hot charcoal seals the juices inside the chicken pieces.

- 2 pounds chicken legs and/or breasts
- 1/4 cup extra virgin olive oil
- 2 tablespoons dry oregano
- 1 lemon, juice of
- 1/4 cup white wine
- 1 medium onion, chopped
- Salt, pepper
- 2 tablespoons fresh thyme, chopped

Rinse chicken thoroughly with water and drain.

In a large mixing bowl, combine oil, oregano, lemon juice, wine, and onion. Add chicken pieces and mix well to coat chicken with marinade. Refrigerate for 3 hours, or for better results, overnight.

Grill over hot coal or gas grill until golden. While grilling, brush chicken frequently with marinade juices.

Remove from grill and sprinkle with salt and pepper. Place onto a serving platter and garnish with thyme. Serve with green salad and oven-baked potatoes.

Preparation time: 15 min.   Cooking time: 30 min.   Yield: 4 servings

Chicken with Lemon
(Kotopoulo Lemonato)

This is a simple, but very delicate and refreshing dish. Excellent for any occasion.

4 chicken legs, whole
Salt, pepper
1 tablespoon flour
3 tablespoons extra virgin olive oil
1 tablespoon lemon zest
2 lemons, juice of

Rinse chicken thoroughly with water and drain; sprinkle with salt and pepper and toss in flour.

In a large saucepan, heat oil and sauté chicken. Sauté on both sides until golden brown.

Add lemon zest, lemon juice, 1 cup water, salt, and pepper. Add less lemon, if you prefer.

Cover and simmer over low heat until chicken is tender and juices are absorbed. Remove from heat and serve hot with dandelion salad and oven-baked potatoes or rice pilaf.

Preparation time: 20 min.   Cooking time: 1 hr. 30 min.   Yield: 4 servings

Chicken with Peas
(*Kotopoulo me Araka*)

This dish is popular in early spring when fresh peas are available. Chicken is sautéed in virgin olive oil and then cooked with fresh peas and garnished with fresh dill.

    2 pounds chicken breasts and/or legs
    1 pound fresh peas
    Salt, pepper
    1 tablespoon flour
    1/4 cup virgin olive oil
    1 medium onion, finely chopped
    4 scallions, chopped
    3 tablespoons fresh dill, snipped
    1 cup thin Avgolemono sauce (optional), see page 91

Rinse peas with water and set aside to drain.

Rinse chicken thoroughly with cold water and drain. Sprinkle with salt and pepper; toss in flour and set aside.

In a large saucepan, heat oil and sauté chicken on both sides until golden. Remove chicken and set aside.

In the same oil, add onions and scallions; sauté until soft (about 3-4 minutes).

Add chicken, 1 cup water, salt, and pepper; cover and simmer over low heat until chicken is half cooked (about 40-50 minutes).

Add peas, 2 tablespoons of the dill, cover, and simmer an additional 30 minutes until peas are tender, but not overcooked.

When cooked, uncover and simmer until water evaporates and the food is left with its oil only. Remove from heat and set aside.

Using water, make Avgolemono sauce, and pour into saucepan. Shake saucepan to coat the chicken and the peas with the Avgolemono sauce. Arrange on a serving platter and garnish with dill. Serve hot.

Preparation time: 20 min.   Cooking time: 1 hr. 30 min.   Yield: 4 servings

Roasted Chicken with Potatoes
(Kotopoulo ston Fourno me Patates)

This dish can be served year-round and for any occasion. The chicken is rubbed with fresh sage and garlic and baked until tender. Potatoes are baked in the chicken juices to enhance the flavor.

> 1 (3 pounds) whole chicken
> 2 tablespoons fresh sage, finely chopped
> 1 small onion, peeled
> 1 celery stalk, cut in small pieces
> 2 garlic cloves, whole
> 1/4 cup extra virgin olive oil
> 3 garlic gloves, crushed
> Salt, pepper
> 4–5 medium potatoes, peeled and quartered
> 1 tablespoon dry oregano
> 1 tablespoon paprika

Preheat oven to 375° F.

Rinse chicken thoroughly with water and drain. Keep skin intact.

Place chicken into a 12 x 9 x 2-inch baking pan with breast up. With your fingers, force half of the sage between the skin and the flesh over the breasts. Place remaining sage, onion, celery, and whole garlic cloves in the cavity.

In a large mixing bowl, combine oil and crushed garlic cloves; brush chicken with oil mixture and sprinkle with salt and pepper.

In the remaining oil mixture, add potatoes, oregano, salt, and pepper; mix well to coat potatoes. Arrange potatoes around chicken and sprinkle with paprika.

Bake for about 90 minutes or until chicken is cooked.

Remove from oven and serve with dandelion salad or carrot salad.

Preparation time: 20 min.   Cooking time: 1 hr. 30 min.   Yield: 4–5 servings

Partridge with Tomato Sauce
(Perdika Salmi)

This is a very popular game dish. It is very difficult to find wild partridge; therefore, when you do, you will want the dish to be absolutely delicious.

    2 partridges, cleaned
    1/4 cup extra virgin olive oil
    Salt, pepper
    3 tablespoons tomato puree
    3 cups red wine
    3 bay leaves
    4–5 strings parsley
    1 tablespoon flour
    2 cups white button mushrooms
    1 teaspoon cornstarch (if required), dissolved in 3 tablespoons water

Clean partridge and with a torch or matches, burn off any small feathers or hairs.

Wash well and set aside to drain. If partridge is split, use a string to tie the bird so that it looks like a whole bird. Preheat oven to 375° F.

Brush partridge with oil, sprinkle with salt and pepper, and place into a small baking dish. Bake for about 25 minutes until half cooked.

Remove from the oven and let cool. Reserve fat from baking dish.

Divide partridge into quarters so that you have 2 breasts with wings, and 2 legs with thighs from each partridge.

In a medium saucepan, heat remaining oil, add tomato puree, and cook for 2 minutes, mixing frequently.

Add partridge, wine, bay leaves, parsley, salt, and pepper; cover and simmer until tender (about 40 minutes). Partridge will not be fully cooked. With a perforated spoon, remove partridge pieces and set aside. Reserve sauce.

In the meantime, pour reserved fat into a skillet and heat; add flour and stir frequently until golden. Slowly add about 1 cup of reserved sauce, stirring vigorously to break up all lumps. Return sauce from skillet into the saucepan and mix. If you have lumps,

Patridge with Tomato Sauce (con't)

pass the sauce through a sieve. Return partridge to saucepan; add mushrooms and mix. Cover and simmer for about 15 minutes or until partridge is thoroughly cooked. Remove from heat; arrange partridge on a serving dish and place mushrooms around partridge.

If sauce is thin, add cornstarch and slowly bring to a boil until sauce thickens.

Pass thick sauce through sieve and pour over partridge.

Serve hot with rice pilaf or mashed potatoes.

Preparation time: 40 min.   Cooking time: 1 hr. 20 min.   Yield: 4 servings

Woodcock with Tomato Sauce (Bekatsa me Domata)

This is another popular game bird. This may be difficult to find, as it is not available as a farm-raised bird. However, the taste is excellent. The birds are tender and have a mild, gamy flavor. Woodcock is about half the size of a partridge.

    4 woodcocks
    Salt, pepper
    1 tablespoon flour
    3 tablespoons extra virgin olive oil
    1 medium onion, grated
    1/2 cup red wine
    8 ounces tomato sauce
    2 bay leaves
    4 slices bread
    2 tablespoons fresh basil leaves, coarsely chopped
    1 teaspoon cornstarch, dissolved in 2 tablespoons water (if required)

Clean woodcock and with torch or matches, burn off any small feathers or hairs. Reserve gizzards.

Split woodcock in quarters, rinse with water, and pat dry. Sprinkle with salt, pepper, toss in flour, and set aside.

On a chopping board, dice gizzards and set aside.

In a medium saucepan, heat 2 tablespoons of the oil and sauté woodcock on both sides until golden brown. With a perforated spoon, remove woodcock pieces from heat and set aside.

Add onions and sauté for 2 minutes. Add gizzards and brown for 2 minutes, mixing frequently.

Return woodcock pieces to saucepan. Add wine, tomato sauce, bay leaves, 1/2 cup water, salt, and pepper; cover and simmer until woodcock is cooked (about 60 minutes).

In the meantime, toast bread until very crispy, but not burned. Brush toasted bread with oil and rub with basil leaves.

Arrange toasted bread slices onto individual serving dishes. Place woodcock pieces on top of the bread.

Pass tomato sauce through a sieve and pour over woodcock. Discard solids. If sauce is too thin, add cornstarch and bring to a boil until sauce thickens. Sauce should have the consistency of pancake batter. Sprinkle with basil leaves.

Serve hot with plain spaghetti or rice pilaf.

Preparation time: 30 min.   Cooking time: 1 hr. 20 min.   Yield: 4 servings

Rabbit with Onions
(Lagos Stifado)

This is a rare dish today. It is very difficult, if not impossible, to find wild rabbit. However, this recipe is also good for farm-raised rabbits. An excellent dish for special occasions. The secret to this dish is that it is cooked slowly and seasoned with aromatic cumin.

    1 (3–4 pound) rabbit
    2 pounds small white onions, frozen
    1/4 cup red wine vinegar
    3 tablespoons flour
    1/4 cup extra virgin olive oil
    2 cups white wine
    1 can (14 ounces) tomato puree
    2 bay leaves
    2 teaspoons ground cumin
    1 cinnamon stick
    Salt, pepper

Cut rabbit into small pieces or ask your butcher to do it for you. Rinse thoroughly with cold water and drain. Dice liver and heart and set aside.

Thaw onions and set aside to drain.

In a mixing bowl, pour vinegar and add rabbit pieces. Stir to coat with vinegar and refrigerate for 2 hours. Drain rabbit and toss in flour.

In a large saucepan, heat oil and sauté onions until soft.

Add rabbit and brown on both sides until golden. Remove rabbit pieces with a perforated spoon and set aside.

Add liver, heart, and brown, mixing frequently. Add remaining flour and brown for 2 minutes, mixing frequently.

Return rabbit pieces to saucepan. Add wine, small onions, tomato puree, bay leaves, cumin, cinnamon stick, 1/2-cup water, salt, and pepper; cover and simmer until onions are soft and rabbit is tender.

To serve this dish, arrange the rabbit pieces in the center of a serving platter and place the onions around the rabbit. Remove and discard cinnamon stick and bay leaves. If sauce is too thin, return to heat and simmer uncovered until juices evaporate and sauce thickens.

Serve with green salad and rice pilaf.

Preparation time: 30 min.   Cooking time: 1 hr. 20 min.   Yield: 4–6 servings

MEAT

No Greek meal is complete without a meat dish. Although lamb is the most popular, beef and pork are also common in everyday cooking.

Let me begin with lamb. Greek food and lamb are synonymous. Spring lamb is used for roasts, Souvlaki, or for grilling. Older lamb or sheep is used in casserole dishes and those made out of ground meat. Stuffed shoulder of spring lamb is delicious. This is a traditional meal for Christmas Day on the island of Mytilini. Ground meat from a sheep will add the aroma and the flavor required for stuffed tomatoes or grapevine leaves.

Easter Sunday is the day when Greeks really show their love for spring lamb. Whole lambs are roasted on large rotisseries over charcoal fires. People celebrate their religious beliefs and the arrival of spring. As lambs are roasted, families and friends gather around and enjoy mezedes from their rich Greek cooking heritage.

Pork is more of a winter meat. Pork chops are eaten with sour orange nerantzi juice and green salad. Pork is also cooked with winter vegetables like celery or cabbage. Pork meat is also used to make excellent Greek sausages.

Beef is also popular. It is not as tender as the beef available in the United States, but the meat is sweeter and less fatty. It is difficult to find 1-inch-thick steaks in Greece. The majority of the steaks are less than 1/2-inch thick and steaks are eaten well-done.

Grilled meats like steaks, pork chops, or lamb chops are served with oregano and lemon sauce. It is an excellent way to enjoy meat in Greece.

Layered Eggplants with Meat Sauce (Mousaka)

Mousaka is, without question, the most famous Greek dish. You will find it in every Greek restaurant and bazaar. It takes a while to prepare, but is worth the effort. I prefer ground lamb meat because it is flavorful, but you can also use ground beef.

    1 pound ground meat (beef or lamb)
    2 pounds (3–4 medium size) eggplants
    1 cup virgin olive oil
    2 medium onions, grated
    1/4 cup red wine
    2 tablespoons tomato paste, dissolved in 4 tablespoons water
    8 ounces tomato sauce
    2 bay leaves
    2 tablespoons parsley, finely chopped
    1 teaspoon sugar
    1 teaspoon allspice
    2 tablespoons dry bread crumbs
    2 egg whites (use egg yolks to make Béchamel sauce)
    4 ounces Kefalograviera cheese, grated
    Salt, pepper
    2 tablespoons Graviera cheese, grated
    8 cups Béchamel sauce, see page 95

Slice eggplants 1/2-inch thick, sprinkle with salt. Set aside to drain for 30 minutes.

In a large, deep skillet, heat one-fourth of the oil and sauté onions for 2–3 minutes. Add ground meat and brown for about 15 minutes, mixing frequently.

Add wine, tomato paste, tomato sauce, bay leaves, parsley, sugar, allspice, and 1 cup of water; simmer until liquid is absorbed. Remove from heat.

Stir half the breadcrumbs into meat sauce. Let cool a few minutes then add egg whites. Mix and set aside. Rinse eggplant with water, squeeze to remove excess water to prevent eggplant from absorbing oil when fried. Drain and dry on a paper towels.

Layered Eggplants (con't)

In a large frying pan, heat half the oil and fry eggplants on both sides until golden. Eggplant does not have to be fully cooked, as it will cook in the oven. Place eggplant pieces on a paper towel to drain excess oil. Spread remaining bread crumbs on the bottom of 11 x 7 x 2-inch baking pan.

Arrange half of the eggplants on the bottom of the pan. Add meat sauce and spread evenly over the eggplants. Spread half the Kefalograviera cheese on top of the meat sauce.

Place remaining eggplants over the meat sauce. Spread remaining Kefalograviera cheese evenly over eggplants. Preheat oven to 375° F.

Make Béchamel sauce and let cool for 5 minutes. Beat egg yolks and stir into sauce. Add salt, pepper, nutmeg; mix well. Sauce should be thick so that it will stay on top of the eggplant and not run to the edges or between the pieces. Pour sauce over eggplants. Sprinkle with Graviera cheese and bake on middle oven rack for 60 minutes or until the top is golden. Remove from heat.

Cool 20 minutes to make cutting and removal easier. Cut into square pieces. First piece may be difficult to remove. Use a flat spatula to remove remaining pieces.

Tip: Substitute Swiss cheese for Kefalograviera cheese, cheddar cheese for Graviera and zucchini for the eggplants. Slice zucchini into 1/2-inch rounds; fry in oil or grill. If baking pan is full, place aluminum foil under the pan to protect your oven from spills.

Preparation time: 90 min.   Cooking time: 1 hr.   Yield: 8–10 servings

Spaghetti with Meat Sauce
(Makaronia me Kima)

Who does not like spaghetti with meat sauce? This is a very common Greek dish that is very easy to make. Fresh basil leaves give this dish a wonderful aroma.

1 pound ground beef
1 pound spaghetti #9
1/2 cup extra virgin olive oil
1 medium onion, finely chopped
2 cloves garlic, crushed and chopped
1 (14 ounces) can tomato puree
1 stick cinnamon
2 bay leaves
Salt, pepper
4 tablespoons Graviera cheese, grated
3-4 fresh basil leaves

In a large saucepan, heat half the oil and sauté onions and garlic for 2 minutes, mixing frequently. Add ground beef and brown for 5 minutes, mixing frequently. Add tomato puree, cinnamon, bay leaves, salt, pepper and 1/2 cup water.

Cover and simmer until ground beef is tender and juices are absorbed. Discard cinnamon stick and bay leaves. Keep saucepan on a very low heat to keep it warm. In the meantime, in a large saucepan, bring about 1 liter water to a boil.

Add 2 tablespoons of the oil and then add the spaghetti; simmer until spaghetti is cooked. Remove from heat, rinse with cold water, and drain in a colander.

Wipe saucepan clean, add remaining oil, and heat until very hot, almost smoking. Add spaghetti and toss a few times to coat with hot oil.

Place spaghetti into a large serving bowl and pour meat sauce over spaghetti. Sprinkle with cheese and basil leaves. Serve hot.

Tip: Cook spaghetti to suit your taste. If you prefer, you may skip rinsing the spaghetti with water. Substitute ground beef with ground pork or ground lamb, or any combination. Substitute Jarlsberg cheese for Graviera cheese.

Preparation time: 20 min.   Cooking time: 1 hr.   Yield: 5–6 servings

Spaghetti Layered with Meat and Béchamel Sauce (Pastitsio)

This is another famous Greek recipe. There are layers of spaghetti and ground meat covered with thick Béchamel sauce and baked in the oven. This dish is very popular with all ages, especially with children.

    1 pound ground meat (beef, pork or lamb)
    1 pound spaghetti #8
    3 tablespoons virgin olive oil
    1 medium onion, grated
    1 can (14 ounces) diced tomatoes
    2 tablespoons tomato paste, dissolved in 4 tablespoons water
    2 bay leaves
    1 cinnamon stick
    3 tablespoons parsley, finely chopped
    1 teaspoon sugar
    1 teaspoon ground cloves
    1/4 cup white wine
    Salt, pepper
    1 tablespoon bread crumbs
    2 egg whites (use egg yolks to make Béchamel sauce)
    8 ounces Kefalograviera cheese, grated
    2 tablespoons Graviera cheese, grated
    8 cups Béchamel sauce, see page 95

In a tall saucepan, bring 1 liter water to a boil. Add salt and spaghetti and cook until soft. Drain, wash with water, drain again and set aside.

In a large saucepan, heat oil and sauté onions for 2 minutes. Add ground meat and brown for 5 minutes, mixing frequently.

Add tomatoes, tomato paste, bay leaves, cinnamon, parsley, sugar, cloves, wine, salt, pepper, 1 cup water and simmer. When juices are absorbed, remove from heat and let cool for a few minutes; add bread crumbs, egg whites, and mix well. Remove and discard bay leaves and cinnamon stick. Mix half of the cheese with the spaghetti. In a 12 x 9 x 2-inch baking pan, spread half of the spaghetti.

Add meat sauce. Spread evenly over the spaghetti. Sprinkle with half of remaining cheese. Add the remaining spaghetti and spread evenly over the meat sauce.

Sprinkle with the remaining cheese and set aside.

Preheat oven to 375° F.

Now prepare Béchamel sauce. Remove from heat and let cool for 5 minutes. Sauce should be thick so it will stay on top of the spaghetti and will not run through.

Beat egg yolks and stir into Béchamel sauce. Spread sauce evenly on top of the spaghetti. Sprinkle with Graviera cheese and bake on middle oven rack for 60 minutes, or until top is golden.

Remove from oven and cool for 10 minutes to make cutting and serving easier. Use knife to cut into 3 x 3-inch pieces. First piece may be difficult to remove. Use flat spatula to remove remaining pieces.

Serve with green salad.

Tip: Substitute Parmesan cheese for Kefalograviera cheese and yellow cheddar cheese for Graviera cheese.

Preparation time: 40 min.   Cooking time: 1 hr.   Yield: 12 servings

Beef Cooked in Wine
(Beef Kapama)

This is a very delicious recipe. It is also a versatile recipe that can be served with rice pilaf, spaghetti, or mashed potatoes.

2 pounds beef (shoulder cut)
1 tablespoon flour
2 tablespoons virgin olive oil
1 carrot, chopped
1 small onion, chopped
1 celery stalk, chopped
4-5 strings fresh parsley
2 bay leaves
1/4-teaspoon ground cloves
1 cup white wine
4 tablespoons tomato sauce
Salt, pepper

Cut meat into 8 equal pieces and toss in flour.

In a large saucepan, heat oil and sauté beef on all sides until brown. Remove meat from saucepan and set aside.

In the same oil, sauté carrots, onion, and celery for 2 minutes, mixing frequently.

Add parsley, bay leaves, cloves, wine, tomato sauce, salt, and pepper.

Return meat to saucepan. Add 1cup water, cover and simmer until meat is tender. When meat is cooked, uncover saucepan and simmer to evaporate any excess juices until sauce is thick.

Remove meat from saucepan and place onto a serving dish. Using a sieve, strain vegetables and juices and pour over meat. Discard vegetables.

Preparation time: 30 min.   Cooking time: 2 hrs.   Yield: 4 servings

Stuffed Tomatoes with Meat
(Domates Gemistes)

This is an excellent dish year-round, but it's especially delicious prepared in the summer when tomatoes are at their best. I prefer to use ground lamb for the unique flavor, but ground beef will also work well.

1/2 pound ground lamb
10 medium tomatoes (3 inches diameter)
2 medium onions, finely chopped
1/2 cup extra virgin olive oil
1 teaspoon sugar
1/2 cup dry bread crumbs
1/3 cup long grain rice
4 tablespoons parsley, finely chopped
3 tablespoons fresh mint, finely chopped
Salt, pepper
5 tablespoons Kefalograviera cheese, grated

Wash tomatoes and make a partial slice at the bottom of the tomato (opposite end from stem), without detaching it. Flip over and carefully scoop out pulp with a small spoon. Try not to break the skin. Discard seeds. Dice pulp, place into a medium mixing bowl, and set aside. In the meantime, in a medium saucepan, bring 2 cups water to a boil and blanch onions for 1 minute. Drain onions.

Wipe saucepan clean and heat 2 tablespoons of the oil and sauté onions until soft. Remove from heat, add sugar, and mix. Set aside to cool.

In a medium mixing bowl, add meat, reserved tomato pulp, reserved onions, remaining oil, bread crumbs, rice, parsley, mint, salt, pepper, and mix well. Set aside.

Preheat oven to 350° F.

Drain any water from the tomatoes and with a spoon, stuff them loosely to about three-quarters full. This will allow room for rice to expand.

Arrange tomatoes in a 12 x 9 x 2-inch baking pan with opening on top and the tomato lid open. Sprinkle cheese into the tomatoes and close the lids. Bake for about 90 minutes on the middle oven rack, basting periodically with their own juices.

Stuffed Tomatoes (Con't)

Remove from heat and serve hot. Just before serving, baste tomatoes using 1 to 2 tablespoons of the juices left in the baking pan. Serve with oven-baked potatoes.

Tip: Tomatoes should be ripe and firm. Initially, you do not need water, as tomatoes will provide juices, but periodically check to ensure that there is enough water for the rice to cook. When basting, try to pour the juices inside the tomato without removing the lid. Substitute Parmesan cheese for Kefalograviera cheese.

Preparation time: 40 min.   Cooking time: 1 hr. 30 min.   Yield: 5 servings

Stuffed Tomatoes.

Beef with Onions
(*Kreas Stifado*)

If you like onions, this is the recipe for you. The ground cumin gives a very distinctive flavor to this dish.

    2 pounds beef (chuck or blade)
    3 tablespoons virgin olive oil
    1 medium onion, chopped
    2 garlic cloves, crushed
    1 tablespoon tomato paste, dissolved in 4 tablespoons water
    1/2 cup red wine
    8 ounces can diced tomatoes
    3 bay leaves
    Salt, pepper
    1 tablespoon ground cumin
    1 teaspoon ground allspice
    1 spice bag
    2 pounds small white onions, thawed

Cut meat into 8 equal-size pieces and remove visible fat. In a large saucepan, heat oil and brown beef on all sides. Add chopped onion and garlic; sauté until onion is soft. Add tomato paste and sauté for 1 minute.

Add wine, tomatoes, bay leaves, salt, pepper, and 1 cup water. Place cumin and allspice in the spice bag and add to saucepan. Cover and simmer over low heat for approximately 60 minutes.

Add small white onions, mix, cover, and simmer over low heat until onions are soft. When onions are soft, remove cover and simmer until most of the juices are absorbed.

Remove from heat and discard bay leaves and spice bag.

Serve hot with French fries.

Tip: You may substitute frozen onions for small fresh onions. Fresh onions cook in less time than the frozen ones, so meat should be soft before you add the onions. If you are using frozen onions, remove the outer layer and add the onions after the meat is half cooked. The outer layer of frozen onions is sometimes hard.

Preparation time: 40 min.   Cooking time: 2 hrs.   Yield: 4–6 servings

Meatballs with Garlic and Tomato Sauce (*Souzoukakia Smyrneika*)

I believe the combination of the tomato sauce and garlic in this dish is excellent. The tomato sauce becomes sweet when baked in the oven and adds more flavor to the dish. These are excellent as a starter or as a main course.

    1 pound ground beef
    4 ounces dry bread crumbs
    1/2 cup red wine
    1 egg
    1 tablespoon ground cumin
    2 medium garlic cloves, finely chopped
    Salt, pepper

For Frying
    1/2 cup olive oil
    1 cup flour
    4 cups tomato sauce, see page 96

Prepare tomato sauce and set aside. Sauce should be thin. In a medium mixing bowl, soak bread crumbs in the wine. Add ground beef, egg, cumin, garlic, 1/2 cup water, salt, and pepper; mix well. Shape mixture into oval-shaped balls about the size of a small egg, and toss in flour.

Preheat oven to 400° F.

In a frying pan, heat oil. Brown meatballs 1 minute on each side.

Arrange meatballs in a 12 x 9 x 2-inch baking pan. Cover meatballs with tomato sauce. Cover pan with aluminum foil. Make a few small-sized holes in the foil to allow steam to escape. Bake at 400° F for 30 minutes.

After 30 minutes, remove foil and bake an additional 20 minutes. The meatballs are ready when the sauce is thick and the top is browned.

Serve hot with white rice pilaf or mashed potatoes.

Preparation time: 30 min.   Cooking time: 50 min.   Yield: 18–20 pieces

Summer Vegetables Stuffed with Lamb (Gemista)

I consider this recipe to be a traditional summer dish on the island of Mytilini. The vegetables are sweet, the mint is fresh and aromatic, and the end product is outstanding.

1/2 pound ground lamb
1 pound (3 small-sized) eggplants
3 medium zucchini
2 green bell peppers
2 red bell peppers
2 medium onions, finely chopped
1/2-cup virgin olive oil
1 teaspoon sugar
1/3 cup long-grain rice
1/3 cup parsley, finely chopped
3 tablespoons fresh mint, finely chopped
Salt, pepper
2 cups thick Avgolemono sauce, see page 91

Cut and remove the ends of the eggplants (end opposite from the stem end), and with a small spoon, scoop out the pulp. Leave stem end intact. Do not cut through the eggplant skin. Keep the ends; you will use them to close the opening after stuffing. Salt inside of eggplants immediately to prevent flesh from discoloring. Set aside for 30 minutes. After 30 minutes, wash eggplants with plenty of water to remove the salt and drain well. Shake to remove excess water.

In the meantime, remove both ends from zucchini; keep one end to use to close the opening after stuffing. With a small spoon, remove zucchini pulp, dice, and set aside. Do not cut through zucchini skin and do not penetrate through the other end. Set aside.

Cut a lid from stem end of the pepper, remove, and discard seeds.

Bring 6 cups of water to a boil in a medium saucepan. Blanch peppers and their lids for 2 minutes. Remove peppers; cool immediately in ice water. In the same boiling water, blanch onions for 1 minute. Remove onions. Drain, set aside to cool.

Summer Vegetables (con't)

Wipe saucepan clean. Heat 2 tablespoons oil and sauté onions until soft. Remove from heat, add sugar, and mix. Set aside to cool.

In a medium mixing bowl, add meat, remaining oil, onions, rice, parsley, mint, salt, and pepper; mix well.

Stuff eggplants, zucchini, and peppers loosely to about three-quarters full. This will allow room for rice to expand. Cut reserved end pieces smaller to fit the openings of the stuffed vegetables and loosely plug ends. Do not push end caps in too far; leave room for rice to expand. Cover peppers with lids.

Arrange stuffed vegetables in a large sauce pan. Add 3-4 cups water, cover and simmer until rice is cooked.

Use juices from sauce pan to make Avgolemono sauce.

Place stuffed vegetables onto a large serving dish, pour Avgolemono sauce over vegetables, and serve hot.

Preparation time: 50 min.   Cooking time: 1 hr.   Yield: 4–5 servings

Stuffed Grapevine Leaves
(Dolmadakia me Kima)

This is my favorite Greek dish. I could eat it every day and never tire of it. I love stuffed leaves more when they are made from fresh and tender grapevine leaves, which are available in late spring and during summer.

    1 pound ground lamb or beef
    1 (net weight, drained, 16 ounces) jar, grapevine leaves.
    1/2 cup virgin olive oil
    2 medium onions, finely chopped
    1 teaspoon sugar
    1/3 cup rice
    1/3 cup parsley, finely chopped
    2 tablespoons fresh mint, finely chopped
    Salt, pepper
    2 cups thick Avgolemono sauce, see page 91

Open jar, remove half of the leaves. In a medium saucepan, bring 5–6 cups water to a boil. Gently unroll leaves, place into boiling water, and simmer for 15 minutes. Do not overcook leaves, as you want them to be soft, but firm, when you stuff them. Remove leaves with perforated spoon and place in a colander to drain and cool.

In the same saucepan, bring 3 cups of water to a boil, and blanch onions for 1 minute. Remove onions, drain, and set aside.

Wipe saucepan clean. Heat 3 tablespoons of the oil and sauté onions until soft. Remove from heat, add sugar, and mix. Set aside to cool.

In a medium mixing bowl, add ground meat, half of the remaining oil, rice, reserved onions, parsley, mint, salt, and pepper; mix well.

Line the bottom of a large saucepan with one layer of grapevine leaves.

To stuff the leaves, take one leaf at a time, stem end towards you, rough side up, and remove stem. Place 1/2 tablespoon of meat mixture on the leaf at the stem end. Fold stem end towards rice mixture half a turn. Then fold left side towards the right and the right side towards the left. Then roll leaf to the end.

Arrange stuffed leaves in the saucepan on top of the layer of leaves lining the bottom

Stuffed Grapevine Leaves (Con't)

of the pan. Place the stuffed leaves into rows close together, with the open ends at the bottom. When finished with one layer, start a new one.

Add enough water to the saucepan to cover the leaves. Pour the water slowly into the pan from the side. Add remaining oil. Place a heat-resistant plate on top of the stuffed leaves. Cover and simmer over a low heat for about 1 hour. Remove from heat and let cool.

In the meantime, using the juices from the saucepan, make Avgolemono sauce. Arrange stuffed leaves on a platter, pour sauce over the leaves and serve.

Tip: Grapevine leaves should be tender. If leaves are large, split them in half and remove stem. Dolmadakia should be the size of your thumb. Roll leaves loosely to allow room for rice to expand. Substitute pre-packed leaves with fresh, tender leaves from a grapevine bush. Pre-packed grapevine leaves are available in supermarkets. You will need to squeeze the leaves in order to remove from jar.

Preparation time: 1 hr. 30 min.   Cooking time: 1 hr.   Yield: 30–40 Dolmadakia

Stuffed Cabbage Leaves
(Lahanodolmades)

This is a winter dish. I prefer to prepare this dish using ground lamb or pork, as these meats are very flavorful. However, this dish is also very good when prepared with ground beef.

1 pound ground lamb or pork
3–4 pounds green cabbage (1 or 2 heads)
2 large onions, grated
1/3 cup long-grain rice
1/2 cup virgin olive oil
1/2 cup parsley, finely chopped
1 red bell pepper, diced, seeds removed
Salt, pepper
1 small celery root, peeled and sliced
2 tablespoons fresh dill, chopped
2 cups thick Avgolemono sauce (optional), see page 91

Remove the core of the cabbage by inserting a small knife into it at an angle and rotating the knife until the core is freed.

In a large saucepan, add enough water to cover the cabbage head. Cover saucepan and boil until the outer layer of the leaves become soft. Remove that layer and continue boiling until the next layer is soft. Set leaves into a colander to drain and cool.

Continue boiling and removing soft leaves until all the leaves are soft. Do not overcook as the leaves will become too soft and will break. Reserve core. The core of the cabbage head is not suitable for stuffing, but it is excellent for salad.

In a medium mixing bowl, combine ground meat, onions, rice, half the oil, parsley, bell pepper, salt, and pepper; mix well.

On a flat surface, place one leaf with stem towards you; inside should be on top. With a small knife, remove the end of the hard stem. Place 1 full tablespoon of the stuffing onto the leaf at the stem end.

Stuffed Cabbage Leaves (Con't)

Roll leaf once towards rice mixture; fold the left side towards the right and the right side towards the left. Then roll leaf to the end. Place rolled cabbage leaves into a large saucepan side-by-side, building up into layers.

When completed, fill saucepan with water to cover stuffed cabbage leaves and add remaining oil. Arrange celery slices on top of the stuffed cabbage leaves. Place a small, heat-resistant dish on top of the cabbage leaves to hold them in place. Cover and simmer for about 60 minutes.

To make the Avgolemono sauce, use the broth from the cabbage. If you do not have enough broth, supplement by adding water.

Arrange cabbage leaves on a serving platter or individual dishes and pour Avgolemono sauce over the cabbage. Garnish with dill. Make salad with the cooked cabbage core and serve with the stuffed cabbage leaves.

Tip: Discard the first couple of cabbage leaves, as they are too soft and will break when you try to roll them. Serve the boiled cabbage salad with oil and lemon.

Preparation time: 60 min.   Cooking time: 1 hr.   Yield: 24–28 Dolmades

Lamb with Zucchini
(Arni me Kolokythia)

Lamb is fantastic for this dish. It is an easy recipe to make and the combination of lamb, zucchini, and potatoes is excellent.

    2 pounds lamb, shoulder cut
    2 tablespoons flour
    1/2 cup virgin olive oil
    1 medium onion, coarsely chopped
    3 garlic cloves, finely chopped
    2 fresh ripe tomatoes, seeds removed, diced
    2 tablespoons tomato paste, dissolved in 4 tablespoons water
    1 tablespoon fresh rosemary, finely chopped
    Salt and pepper
    2 medium potatoes, quartered
    5 medium zucchini, each cut into 5 pieces

Cut meat into 8 pieces and toss in flour.

In a large saucepan, heat half of the oil and sauté lamp chops on both sides until golden (about 3 minutes each side). Remove meat from heat and set aside.

In the same saucepan, add onion and garlic, and sauté for 2 minutes.

Add tomatoes, tomato paste, rosemary, salt, pepper, and 1 cup of water. Return meat to saucepan and mix with other ingredients. Cover and simmer for about 60 minutes, until meat is almost tender.

Move meat to the sides of the saucepan to make room in the middle for the potatoes and zucchini.

In the meantime, in a large skillet, heat 2 tablespoons of the remaining oil and brown potatoes for 5 minutes, mixing frequently. Remove potatoes and place in saucepan with lamb.

In the same skillet, brown zucchini for about 3 minutes, mixing frequently. Remove and place into saucepan with meat. Add remaining oil, cover, and simmer until zucchini and potatoes are cooked and juices are absorbed.

Lamb with Zucchini (Con't)

Arrange the meat in the middle of a serving platter or in individual serving dishes, and place potatoes and zucchini around the meat. Baste with juices and serve with green salad.

Tip: The best meat for this dish is shoulder cut lamb or lamb chops. This meat will give more flavor.

Preparation time: 30 min.   Cooking time: 1 hr. 30 min.   Yield: 4 servings

Lamb with Oregano
(Arni me Rigani)

The name of this recipe reveals that the predominant flavor is oregano. Oregano grows everywhere on the island of Mytilini. In May, the oregano plant forms a flower, and by June, the flower dries up. This dried flower is then picked and processed to produce the oregano sold in the marketplace. This is a delicious and very aromatic dish.

    4 lamb knuckles or 2 pounds leg of lamb
    4 garlic cloves, each sliced into 4 pieces
    Salt, pepper
    2 tablespoons flour
    4 tablespoons virgin olive oil
    1/2 cup white wine
    2 tablespoons dry oregano
    2 bay leaves
    2 lemons, juice of
    1 tablespoon fresh oregano, chopped

Remove visible fat from knuckles, wash and dry. If you are using leg of lamb, cut lamb into 4 or 8 equal pieces, and remove visible fat.

With a sharp knife, score the knuckles or the lamb pieces in a few places and fill cuts with garlic slices. Sprinkle meat with salt and pepper and toss in flour.

In a large-bottom saucepan, heat oil and brown meat on all sides. Add wine, oregano, bay leaves, lemon juice, salt, pepper, and 2 cups water. Cover and simmer over low heat until lamb is tender (approx. 2 hours). Add more water, if required.

Remove from heat and discard bay leaves. This dish is served with mashed potatoes.

To serve, arrange mashed potatoes in the center of a serving dish. Place lamb knuckles on top and soak with juices. Sprinkle with fresh oregano and serve. Also, serve green salad with this dish.

Tip: Lamb knuckles were very difficult to find, but in the last two years, they have become available in supermarkets. In fact, most of the lamb knuckles available in the supermarkets are imported from New Zealand.

Preparation time: 30 min.   Cooking time: 2 hrs.   Yield: 4 servings

Roasted Leg of Lamb with Potatoes
(Arni me Patates)

This is the traditional Greek way of cooking lamb in the oven. It is a simple, but very tasty and hearty recipe. Spring lamb is ideal for this dish.

    1 leg of lamb (4 pounds) or 2–3 pounds boneless
    3 pounds potatoes, quartered
    3 garlic cloves, quartered
    1 lemon, juice of
    1/4 cup virgin olive oil
    Salt, pepper
    1 tablespoon paprika
    2 tablespoons fresh rosemary, chopped

Preheat oven to 450° F.

Remove visible fat from meat and score with a sharp knife in several places. Mix garlic and 1 tablespoon of rosemary and fill the cuts of the meat. Boneless lamb is sold in holding nets or tied with string; keep it tied. Place lamb in the middle of a 15 x 12 x 2-inch roasting pan.

In a large mixing bowl, combine lemon juice, remaining rosemary, and oil. Add potatoes and mix well to coat all around.

Arrange potatoes in the baking pan around the lamb. Brush meat with the oil mixture and pour the remaining mixture over the potatoes.

Sprinkle lamb and potatoes with salt, pepper, and paprika.

Roast at 450° F for about 30 minutes and then reduce heat to 350° F and roast for an additional 90 minutes. When cooked, remove from oven and let it rest for 10 minutes before carving.

To serve this dish, arrange potatoes in the middle of a serving platter. Slice meat into thin pieces and arrange around the potatoes.

Discard excess fat from the pan and pour juices over lamb.

Serve with dandelion salad or green salad.

Tip: Remove visible fat, avoiding cutting through the meat. The roasting time shown above will produce a well-done leg of lamb. Therefore, if you prefer your roast medium or rare, you should use a thermometer and follow the Temperature Tables on page 14.

Preparation time: 20 min.   Cooking time: 2 hrs.   Yield: 6–8 servings

Roast Leg of Lamb with Potatoes.

Roasted Lamb in a Bag
(Arni stin Sakoula)

Roasting lamb in a bag allows the lamb to cook in its own juices. It takes longer to cook, but it is worth the effort, as the meat will be very juicy and tender.

    1 leg of lamb (4–5 pounds) or 2–3 pounds boneless
    3 garlic cloves, sliced
    3 tablespoons fresh rosemary, finely chopped
    2 tablespoons mustard paste
    2 pounds small potatoes, whole
    3 tablespoons tomato puree
    Salt, pepper
    Cotton string
    Parchment paper (about 18 inches wide)
    1 teaspoon cornstarch, dissolved in 3 tablespoons cold water (optional)
    2 tablespoons fresh oregano, finely chopped

Score the leg in several places with a sharp knife, making deep incisions. Mix garlic and half of the rosemary and fill meat cuts. Then spread mustard evenly over the lamb.

Roll out 3 pieces of parchment paper about 2 feet long and place into a 12 x 9 x 2-inch baking pan.

Place lamb on parchment paper. Add potatoes around the lamb, and sprinkle with remaining rosemary, tomato puree, salt, and pepper.

Wrap tightly with parchment paper and tie with string to keep paper together. Opening should be on top to prevent juices from seeping out.

Roast in a 400° F oven for approximately 3 hours. Remove from oven and set aside to cool for 5–10 minutes.

Cut string to open parchment paper. Lamb meat should be tender enough to be able to be cut with a fork.

To serve, place meat in the middle of a serving platter, arrange potatoes around the meat, and baste with a couple tablespoons of the juices. Skim fat from the top before using the juices.

If you prefer a thick sauce, then pour about 1 cup of the juices into a small saucepan, add the dissolved cornstarch, and slowly bring to a boil until sauce thickens. Sprinkle potatoes and lamb with oregano. Serve with dandelion salad.

Tip: You can use a large oven-roasting bag in place of the parchment paper and bake only 2 hours. Tie the end and do not make holes in the bag.

Preparation time: 30 min.   Cooking time: 3 hrs.   Yield: 6–8 servings

Stuffed Shoulder of Lamb
(Arni Gemisto)

A traditional Christmas dish on the island of Mytilini; most islanders prepare this dish on Christmas Day and will bake it in the town oven. The town oven is heated with olive tree wood and stays open on Christmas Day just to cook this traditional meal for residents.

  1 lamb shoulder (4–6 pounds)
  Salt, pepper
  1/2 cup virgin olive oil
  1 medium onion, finely chopped
  2 ounces lamb's liver, diced
  2 ounces lamb kidneys (optional), diced
  1/2-cup long-grain rice
  1 tablespoon tomato paste, dissolved in 4 tablespoons water
  2 tablespoons raisins, seedless
  3 tablespoons pine nuts, roasted
  3 tablespoons fresh dill, finely chopped
  6-8 chestnuts, roasted and cut in quarters
  Cotton string and needle
  3 pounds potatoes, peeled and quartered
  1 lemon, juice of
  2 tablespoons fresh rosemary, finely chopped
  1 teaspoon paprika

With a small, sharp knife and your fingers, make a dip pocket between the ribs and flesh through to the shoulder bone. You are making a large pocket. Start from the ribs first. It is important not to cut any holes in the skin to prevent the stuffing from spilling out of the pocket. Sprinkle the meat inside and out with salt and pepper and set aside.

In a medium saucepan, heat 3 tablespoons oil and sauté onions until soft. Add liver, kidneys, and sauté for 2 minutes.

Add rice and sauté for 3 minutes mixing frequently. Then add tomato paste, 1 cup water, raisins, pine nuts, dill, chestnuts, salt, and pepper; mix together.

Cover and simmer over low heat until water is absorbed. Rice will not be fully cooked. Remove from heat and set aside to cool.

Stuff shoulder cavity with rice mixture and tie opening with the string. Place lamb in the middle of an 18-inch-round roasting pan and arrange potatoes around the lamb.

In a small mixing bowl, combine remaining oil with lemon juice and brush lamb and potatoes. Pour remaining mixture over potatoes. Sprinkle lamb and potatoes with rosemary and paprika and roast at 400° F for 30 minutes. Reduce heat to 350° F, cover lamb only (not potatoes) with aluminum foil and roast at 350° F for approximately 90 minutes.

Remove foil and roast lamb an additional 30 minutes, basting occasionally with its own juices, until top of lamb becomes a nice golden color.

To serve, remove string and discard. Remove stuffing and place on one side of a serving platter. Slice meat and arrange on the other side of the platter. Pour some of the juices over the meat. Serve with green salad or dandelion salad.

Tip: Shoulder should be from a young lamb. It is important to let the shoulder cook well, because the neck side of the shoulder takes longer to cook. If you accidentally cut through the flesh, use aluminum foil to block the hole from the inside and remove the foil before serving.

Preparation time: 50 min.   Cooking time: 2 hrs. 30 min.   Yield: 4–6 servings

Lamb with Orzo
(Arni Giouvetsi)

This is another traditional Greek dish. Meat is slowly cooked in tomato sauce and then baked in the oven with the orzo. While lamb makes this dish very flavorful, you may also prepare this delicious entrée with beef (shoulder cut).

    2 pounds leg of lamb, boneless
    2 garlic cloves, sliced
    Salt, pepper
    5 tablespoons virgin olive oil
    1 medium onion, coarsely chopped
    1/2-pound orzo
    1 (8 ounces) can diced tomatoes
    2 bay leaves
    6–8 sun-dried tomatoes, coarsely chopped
    1 large tomato, thinly sliced
    4 ounces Graviera cheese, grated
    2 tablespoons fresh basil leaves, chopped

Remove visible fat from meat and cut into 8 equal pieces.

With a sharp knife, score meat pieces in 2–3 places. Fill cuts with garlic slices and season with salt and pepper.

In a large saucepan, heat oil and brown lamb. Remove lamb and in the same oil, sauté onions. Return lamb to saucepan, add diced tomatoes, bay leaves, salt, pepper, and 2 cups water. Cover and simmer for approximately 2 hours or until meat is tender.

In the meantime, boil 6 cups water in a medium saucepan. Add orzo and simmer until cooked. Remove from heat, drain, and rinse with cold water. Drain again.

In a 12 x 9 x 2-inch baking pan, arrange lamb pieces. Pour orzo into saucepan with tomato sauce, add dried tomatoes and mix. Now pour orzo into baking pan, over the lamb. Try to cover the lamb pieces with orzo. Place tomato slices over lamb pieces. The idea is to keep lamb pieces covered with orzo and tomato pieces to prevent lamb from getting dark when cooked.

Bake in a 400° F oven until all juices are absorbed or until orzo becomes a light golden color (about 30 minutes). The dish is ready when orzo is left almost dry, with very little juices.

Remove from oven and immediately sprinkle with cheese. Discard bay leaves.

Serve hot with tomato salad or green salad. Sprinkle with basil leaves before serving.

Tip: You may substitute sharp cheddar or Parmesan for the Graviera cheese.

Preparation time: 40 min.   Cooking time: 2 hrs. 30 min.   Yield: 6–8 servings

Lamb with Eggplants
(Arní me Melítzanes)

I learned this recipe from my father-in-law. I think it is an excellent dish for special occasions. Meat is cooked in tomato sauce; the eggplants are sautéed and then baked in the oven.

   2 pounds leg of lamb, boneless
   2 pounds or 2 large eggplants
   Salt
   1 cup virgin olive oil
   1 small onion, grated
   2 garlic cloves, finely chopped
   8 ounces tomato sauce
   1/4 cup red wine
   1 teaspoon ground allspice
   Pepper
   2 large tomatoes, thinly sliced

Slice eggplants into 3/4-inch-thick rounds. Discard stems. Sprinkle with plenty of salt and set aside in a bowl for 30 minutes.

After 30 minutes, wash eggplants with water to remove salt and squeeze to remove excess water and air pockets. The salt will take away any bitterness and the squeezing will eliminate the porosity, so that the eggplant will not absorb oil.

In the meantime, cut lamb into 1-inch to 2-inch cubes. In a medium saucepan, heat 2 tablespoons of oil and brown meat, mixing frequently.

Add onion and garlic, and sauté until onion is soft. Add tomato sauce, wine, allspice, salt, pepper, and 1/2 cup water. Cover and simmer over low heat until meat is tender. When meat is tender, set aside. In a large frying pan, heat half of the remaining oil and fry eggplant slices until golden, turning once. Do not overcook eggplants as they will be cooked again in the oven. Place on a paper towel to remove excess oil.

Arrange eggplants side-by-side in a 12 x 9 x 2-inch baking pan. Place 2–3 pieces of the cooked meat on top of each eggplant slice. Pour some of the tomato sauce over the lamb and eggplant slices. Arrange tomato slices over the lamb pieces to prevent it from drying and discoloring when baked.

Cover pan with aluminum foil and bake at 375° F for 40 minutes. After 40 minutes, remove aluminum foil and bake for an additional 20 minutes.

Tip: Substitute fried eggplants with grilled eggplants. For grilled eggplants, you will need only 1/4 cup oil.

Preparation time: 40 min.   Cooking time: 2 hrs. 10 min.   Yield: 6 servings

Lamb Kebab
(Arni Souvlaki)

Who isn't familiar with Souvlaki? Most of the time, Souvlaki is made out of pork meat, but I prefer to use lamb. Souvlaki is great when grilled on a gas or charcoal grill. To enhance the flavors, it is essential to marinate the lamb for a few hours.

    2 pounds leg of lamb, boneless
    1/4-cup olive oil
    1/4-cup red wine
    1 small onion, chopped
    2 tablespoons dry oregano
    8–10 cherry tomatoes
    1 medium onion, quartered, layers separated
    1 green bell pepper, cut into 1-inch squares
    1 lemon, juice of
    Salt, pepper
    4–5 metal skewers, 12 inches long
     Tzatziki sauce (optional), see page 90

Prepare lamb by removing all visible fat and silver muscle. Cut lamb into about 1 1/2-inch cubes and set aside. You will get 18–22 pieces.

To make marinade, in a medium mixing bowl, combine oil, wine, chopped onion, and half of the oregano. Add lamb, mix well. Cover bowl with plastic wrap or foil and refrigerate for 3–4 hours or overnight for better results.

Skewer ingredients, starting with lamb and alternating with cherry tomato, onion layer, and green pepper squares. Start with lamb and finish with lamb to prevent the vegetable pieces from falling off while grilling.

Grill lamb over charcoal, brushing frequently with the marinade. Grill on all sides.

Remove from heat and sprinkle with lemon juice, salt, pepper, and oregano. Serve hot. Serve with oven potatoes, rice pilaf, or just with salad. Tzatziki sauce is an excellent dressing for this dish.

Tip: If available, use flat metal skewers. If you are using wooden skewers, soak them in water for 1 hour before use. This will prevent skewers from burning on the grill.

Preparation time: 40 min.   Cooking time: 15–20 min.   Yield: 4–5 Souvlaki

Lamb with Cauliflower
(Arní me Kounoupídi)

This is not a well-known recipe, but in winter when cauliflower is sweet, it makes an excellent meal.

1 1/2 pounds lamb (shoulder cut)
1 tablespoon flour
4 tablespoons virgin olive oil
1 medium onion, finely chopped
1 teaspoon sugar
6 ounces tomato sauce
Salt, pepper
1 (2–3 pounds) cauliflower head
1 tablespoon fresh oregano, chopped

Cut lamb into 3-inch cubes and lightly toss in flour.

In a large saucepan, heat oil and sauté lamb on all sides. Remove lamb from saucepan and set aside.

In the same oil, sauté onions, mixing frequently.

Return lamb to saucepan; add sugar, tomato sauce, 1 cup water, salt, and pepper. Cover and simmer over a low heat until lamb is soft. Add more water, if required.

In the meantime, break the cauliflower into florets. Wash with water, drain, and when meat is cooked place into saucepan with meat. Mix to coat cauliflower with the sauce. Cover and simmer for approximately 8-10 minutes or until cauliflower is soft. Do not overcook, as cauliflower will crumble.

Remove from heat and serve in individual dishes. With a large spoon, pour some of the sauce over the lamb and cauliflower.

Sprinkle with fresh oregano.

Preparation time: 30 min.   Cooking time: 1 hr. 50 min.   Yield: 4 servings

Lamb with Okra
(Arni me Bamies)

Okra is another vegetable that goes well with lamb. This is a summer dish, as fresh okra is plentiful in the summertime. Lamb is browned in olive oil and then, baked in the oven with okra.

1 1/2 pounds lamb, lean shoulder cut
2 pounds fresh okra or frozen
1/2 cup red wine vinegar
Salt, pepper
4 tablespoons virgin olive oil
2 medium onions, finely chopped
1 (8 ounces) can crushed tomatoes
3 tablespoons chopped parsley
1 large, ripe tomato, sliced into thin rounds

Thaw frozen okra, rinse with water, and drain well.

If you are using fresh okra, remove cone-shaped ends from the okra without cutting through the pods. Discard ends. Rinse with water, drain, and place into a medium mixing bowl.

Add vinegar to okra, mix together, and set aside for 20 minutes. After 20 minutes, rinse okra with cold water to remove vinegar. Set aside

In the meantime, cut lamb into 3-inch cubes and sprinkle with salt and pepper.

In a large saucepan, heat oil and brown lamb. With perforated spoon, remove lamb and set aside.

In the same oil, add onions and sauté until soft. With perforated spoon, remove onions and place with lamb.

In the same oil, sauté okra. With perforated spoon, remove okra and set aside.

Return lamb and onions to saucepan, add crushed tomatoes and parsley; cover and simmer until lamb is almost cooked.

Remove from heat and arrange meat in a 12 x 9 x 2-inch baking pan and cover with okra. Pour all juices into the saucepan. Mix to coat okra with sauce. Arrange tomato slices so that they cover the lamb pieces. This will keep the lamb from turning dark when cooked in the oven.

Cover with aluminum foil and bake at 375° F for approximately 40 minutes. After 40 minutes remove foil and bake an additional 20 minutes. If using frozen okra, bake uncovered for 1 hour. Sprinkle with freshly ground pepper and serve hot.

Preparation time: 30 min.   Cooking time: 2 hrs.   Yield: 4–5 servings

Lamb with Raspberry Beans
(Arni me Fasolia Barbounia)

This is my mother's recipe. We always had fresh raspberry beans growing in the garden and she would make this dish on special occasions. This dish is very tasty, but very difficult to find. The reason is that fresh raspberry beans are very scarce. However, you can use frozen or canned beans.

  2 pounds lamb chops
  1 pound fresh or frozen raspberry beans, without shells
  1/4 cup extra virgin olive oil
  1 medium onion, finely chopped
  1 tablespoon tomato paste, dissolved in 4 tablespoons water
  1 teaspoon sugar
  1 (8 ounces) can diced tomatoes
  2 tablespoons fresh parsley, finely chopped
  Salt, pepper

If you are using frozen beans, thaw, rinse and drain. If you are using fresh beans, wash and drain.

In a large saucepan, bring to a boil 6 cups water and cook beans for 20 minutes. Remove from heat, drain, and set aside.

Wipe saucepan clean. Heat oil and sauté lamb chops on all sides; add onion and sauté for 2 minutes. Add tomato paste and sauté for 2 minutes.

Return raspberry beans to the saucepan. Add sugar, tomatoes, 1 1/2-cups water, parsley, salt, and pepper; mix well. Cover and simmer over low heat until meat and beans are tender. Add more water, if needed.

Serve hot.

Tip: The reason for removing the shells from the raspberry beans is that the shells are very hard. If you find young, fresh beans, you may be able to use the shells. However, I always recommend that you discard the shells.

Preparation time: 40 min.   Cooking time: 2 hrs.   Yield: 4–5 servings

Pork with Lettuce
(Hirino Frikase)

Lettuce is cooked with pork and covered with Avgolemono sauce. The lemony sauce goes very well with this dish. You must try it to appreciate the flavors.

- 2 pounds pork, shoulder cut
- 1 tablespoon flour
- 4 tablespoons virgin olive oil
- 1 large onion, coarsely chopped
- 1 bunch scallions, chopped
- 1/2-cup white wine
- Salt, pepper
- 3 tablespoons fresh dill, snipped
- 3 pounds romaine lettuce
- 1 cup thick Avgolemono sauce see page 91

Cut pork into 8 equal-size pieces and toss in flour.

In a large saucepan, heat oil and brown pork on all sides. Add onion, scallions, and sauté for 2 minutes.

Add wine, 1 cup water, salt, pepper, and half of the dill. Cover and simmer until pork is tender.

In the meantime, wash lettuce, drain well, and set aside to dry. Keep lettuce leaves whole.

When pork is tender, push it to one side of the saucepan and place the lettuce at the other side of the saucepan. Cover and simmer until lettuce is tender. Do not overcook. Add more water if needed. You should have about 1 cup of broth left in the saucepan when pork is cooked. Use this broth to make the Avgolemono sauce.

Pour Avgolemono sauce in saucepan and shake to cover pork and lettuce.

On individual dishes, place lettuce first, and then place the pork on top of the lettuce. Pour some Avgolemono sauce over the meat.

Sprinkle with freshly ground pepper, remaining dill, and serve.

Preparation time: 20 min.   Cooking time: 1 hr. 30 min.   Yield: 4 servings

DESSERTS

Greeks, like many other people, love their desserts, which are often served in the afternoon after their siesta, or late at night, after a nice dinner with friends.

Filo dough is used to make the famous Baklava, Galatoboureko, or Bougatsa. Extra virgin olive oil is used in many instances, instead of butter. However, most of the recipes in this book are made using butter. Eggs, farina, walnuts, almonds, honey, and flour are the other main ingredients found in most Greek desserts.

Making desserts is a long and tedious process. You must follow the recipe exactly and must use the amounts shown on the ingredient list. You may substitute only the ingredients as indicated in the list. If you follow the instructions patiently, you will achieve excellent results.

Shredded-Dough Walnut Pastry (Kadaifi)

This dessert likes syrup. You should make sure that the shredded dough is baked until crispy and your syrup is thick. I like my kadaifi crispy on top and soft on the bottom.

    1 pound kadaifi dough
    12 ounces walnuts, finely chopped
    1 teaspoon ground cloves
    2 tablespoons sugar
    1 teaspoon ground cinnamon
    1/4-cup bread crumbs, plain
    8 ounces butter, melted, clarified
    1–2 tablespoons rose water (optional)

Syrup
    2 cups water
    2 1/2 cups sugar
    1 stick cinnamon
    2 tablespoons lemon juice

Remove kadaifi dough from refrigerator and keep at room temperature for 3–4 hours. If frozen, thaw at room temperature for at least 10 hours. Keep in original packaging while thawing.

In a medium mixing bowl, combine walnuts, cloves, sugar, cinnamon, and bread crumbs. Melt butter and keep warm.

Preheat oven to 350°F.

Separate kadaifi dough into 12–14 equal parts and cover with a kitchen towel to keep it from drying. Take one kadaifi dough portion at a time and form a rectangular shape about 3 inches wide by about 6–7 inches long. Lay flat on a flat surface, with one end towards you. With a brush butter dough lightly.

Place 1 to 1 1/2 tablespoons of walnut mixture on the end towards you and gently roll the dough to make a roll. Roll should be about 3 inches wide by 2 inches high.

Shredded-Dough Walnut Pastry (Con't)

Arrange rolls side-by-side in a 12 x 9 x 2-inch, buttered, baking pan. Using a spoon, pour the warm, clarified butter over the kadaifi rolls so as to wet all the dough. Bake on middle oven rack for 50–60 minutes or until tops are golden.

In the meantime, make syrup as shown below. It will take about 20 minutes to make the syrup.

Remove from oven and immediately sprinkle with rose water, if using. Then immediately pour the hot syrup over the kadaifi.

Let it cool a few hours to allow the pastry to absorb the syrup. Do not cover.

To Make the Syrup
In a medium saucepan, combine water and sugar, add cinnamon stick, and bring to a boil. Then reduce the heat and simmer about 10–12 minutes or until the syrup is thickened. Add lemon juice just before you remove syrup from the heat. Remove and discard cinnamon stick. Pour over hot kadaifi.

Tip: Kadaifi shredded dough can be found in the frozen section of almost every supermarket, Italian delicatessen, or Greek food store. Dough should be soft, otherwise it will break when you work with it. To achieve a golden brown color without dark spots, you must clarify butter before pouring over the dough.

Preparation time: 30 min.   Baking time: 1 hr.   Yield: 12–14 pieces

Baklava with Almonds
(Baklava me Amygdala)

This delight is found in the town of Papados, on the island of Mytilini. It is rich, but it is simply the best. Baklava needs good butter; therefore, use the best butter available.

 2 pounds filo dough #4
 3 pounds almonds, blanched, finely chopped
 4 tablespoons sugar
 15 pieces bitter almonds, finely chopped (or 1/2 teaspoon bitter almond extract)
 1 tablespoon ground cinnamon
 1/2 cup white bread crumbs, plain
 2 pounds butter, melted, clarified

Syrup
 6 cups water
 5 cups sugar
 1 cinnamon stick
 4 tablespoons lemon juice
 2 ounces butter

Remove filo dough from refrigerator and keep at room temperature at least 6 hours. If frozen, thaw at room temperature for at least 12 hours or overnight. Keep in original packaging while thawing.

Combine almonds, sugar, bitter almonds, cinnamon, and bread crumbs. Set aside. Unroll dough, lay flat, measure and cut dough to fit a 15 x 9 x 2-inch baking pan. Save the trimmings to use between the layers. Brush the sides and bottom of baking pan with butter.

Place 1 filo pastry sheet flat on the bottom of the pan and brush the top with butter. After the eighth sheet, spread a thin layer of the almond mixture to cover the pastry sheet. Add 4 additional sheets, brushing each sheet with butter. Add almond mixture again. Continue alternating between 4 dough sheets and almond mixture until all almonds are used. Finish with 6–8 dough sheets brushing each with butter.

Pour remaining butter on top of dough and place the pan in refrigerator for about 30 minutes. Remove the baklava from the refrigerator and with a sharp knife, cut dough all the way through lengthwise, into 2-inch strips. Brush top with cold water.

Baklava with Almonds (con't)

Preheat oven to 275° F. Bake on middle rack for 2 hours and 30 minutes. Then increase heat to 375° F and bake for 10 more minutes or until top is golden brown.

In the meantime, make the syrup. See below. It takes 30 minutes to make the syrup.

Remove from the oven and immediately pour the cold syrup over the top of the Baklava, ensuring even distribution. Set aside for a few hours to cool and to absorb the syrup. Cut into small pieces and serve.

To Make the Syrup
In a medium saucepan, combine water, sugar, and add cinnamon stick. Bring to a boil, reduce heat and simmer for 25-30 minutes or until syrup thickens. Add lemon juice, butter, and mix to melt butter. Remove from heat to cool. Discard cinnamon stick.

Tip: Filo dough can be found in the frozen section of almost every supermarket, Italian delicatessen, or Greek food stores. Butter must be clarified to eliminate dark spots on top of the filo dough. For Baklava to stay crispy, do not cover.

Preparation time: 50 min.   Cooking time: 2 hrs. 40 min.   Yield: 35–40 pieces

Baklava with Walnuts
(Baklava me Karidia)

This is the most famous traditional Greek dessert. It is a rich dessert with cinnamon and clove flavors.

  1 pound filo dough #4
  1 pound walnuts, finely chopped
  1 teaspoon ground cinnamon
  2 tablespoons sugar
  1 pound butter, melted, clarified
  35-70 cloves, whole

Syrup
  1 cup water
  2 cups sugar
  1 cinnamon stick
  2 tablespoons lemon juice
  2 ounces butter

Remove dough from the refrigerator and set at room temperature at least 6 hours. If frozen, thaw at room temperature for at least 12 hours or overnight. Keep in original packaging while thawing.

In a medium mixing bowl, combine walnuts, cinnamon, and sugar and set aside.

On a flat surface, roll out pastry sheets and cut then to fit a 17 x 12 x 1-inch baking pan. Keep trimmings. Grease the bottom and sides of the baking pan.

Place 1 filo sheet flat in the baking dish and brush the top with butter. Add a second sheet and brush it with butter. Sprinkle some trimmings over each layer. Repeat this process until you use half of the sheets (about 12–13)

Spread walnut mixture evenly over the dough and cover with the remaining sheets, brushing again each sheet with butter. Keep spreading trimmings between the layers. The last pastry sheet should be laid evenly as one piece, as this will represent the final look of the Baklava.

Baklava with Walnuts (con't)

Pour all remaining butter on top and refrigerate for about 30 minutes. Preheat oven to 350° F. In the meantime, make the syrup and let it cool before using.

With a very sharp knife cut dough into square- or triangle-shaped pieces, insert 1 clove on top of each piece and bake on the middle rack for 60 minutes or until top of dough is golden. This size baking pan will make about 35 square pieces or 70 triangles.

Remove from oven and immediately pour cold syrup over. Set aside for a few hours to cool and to let pastry absorb the syrup. Don't cover so Baklava will stay crispy.

To Make the Syrup

Combine water and sugar; add cinnamon stick and bring to a boil. Reduce heat and simmer about 15 minutes or until syrup thickens. Before removing from heat, add lemon juice, butter, and mix to melt butter. Discard cinnamon stick; set aside to cool.

Tip: Substitute walnuts with pistachio nuts. The filo pastry can be found in the frozen section of almost every supermarket, Italian delicatessen or Greek food store. Clarify butter before brushing the top filo sheet to eliminate dark spots. Syrup should be thick and sticky and not watery.

Preparation time: 30 min.   Baking time: 1 hr.   Yield: 35–70 pieces

Custard Pastry
(Galatoboureko)

Custard is encased into filo dough, baked in the oven until filo is crispy, and then, covered with aromatic syrup. It is best when it is eaten warm.

    1 pound filo dough #4
    1 1/2 pints milk
    2 cups sugar
    2–3 strips of lemon peel
    8 ounces fine farina
    1 tablespoon vanilla extract
    5 eggs, beaten
    8 ounces butter, melted, clarified

Syrup
    3/4 cup water
    2 cups sugar
    10 cloves, whole
    1 tablespoon lemon juice

Remove filo dough from refrigerator and set at room temperature for 6 hours. If frozen, thaw at room temperature for at least 12 hours or overnight. Keep in original packaging while thawing.

To prepare cream filling combine milk, sugar, and lemon peel, and bring to a boil; reduce heat and simmer at low heat for 30 minutes, mixing continually. Remove lemon peel and discard. Add farina, stir, and simmer for another 30 minutes, mixing continually. Mixture will be a thin cream, the consistency of thick pancake batter.

Remove from heat, stir in vanilla, and let cream cool for 10 minutes. Stir eggs into cream and incorporate well. It is important that you let the cream cool before you add the eggs.

Place 1 filo sheet flat on a 15 x 9 x 2-inch well-buttered baking dish and brush the top with butter. Add a second sheet and brush it with butter. Repeat this process until you use 8–10 sheets.

Custard Pastry (Con't)

After the 8–10 sheets, evenly spread cream mixture over the filo dough and cover with 6–7 sheets, brushing again each sheet with butter. The last filo sheet should be laid evenly as one piece, as this will represent the final looks of the custard pie.

Pour all remaining butter on top and refrigerate for about 30 minutes. Preheat oven to 325° F. In the meantime, make the syrup and let it cool before using.

Cut the top sheets of the cooled galatoboureko into 3 x 3-inch squares and bake on the middle rack for about 2 hours or until filo dough is golden. Remove from the oven and slowly pour cool syrup over the hot Galatoboureko. Do not cover pan because Galatoboureko will become soggy.

Cut pieces through before serving.

To Make the Syrup
Combine water with sugar; add cloves and bring to a boil. Reduce heat and simmer for about 15 minutes. Stir in lemon juice, remove from heat, and set aside to cool. Remove and discard cloves.

Tip: The filo dough can be found in the frozen section of almost every supermarket, Italian delicatessen, or Greek food store. Wrap and freeze unused filo dough.

Preparation time: 1 hr.   Cooking time: 2 hrs.   Yield: 15 pieces

Walnut Cake
(Karidopita)

This cake is very light and very fluffy. The brandy and spices give this cake a unique and delightful flavor.

    12 ounces walnuts, finely chopped
    8 eggs (separate yolks and whites)
    1 cup sugar
    1 teaspoon vanilla extract
    3 tablespoons brandy
    4 tablespoons bread crumbs, plain
    1 teaspoon ground cinnamon
    1 teaspoon ground cloves
    1 tablespoon baking powder
    2–3 tablespoons confectioners' sugar, (optional, see Tip)

Syrup
    1 cup water
    1 1/2-cup sugar
    1 cinnamon stick
    1 tablespoon lemon juice

Grease a 12 x 9 x 2-inch baking pan and set aside.

Beat egg whites into a soft meringue and set aside. Then beat egg yolks with sugar until color turns pale yellow. Stir in vanilla and brandy and set aside.

Preheat oven to 350° F.

In a large mixing bowl, combine walnuts, bread crumbs, cinnamon, cloves, and baking powder. Add egg yolks and stir.

Fold meringue into walnut mixture and then pour mixture into the baking pan (ensure meringue is incorporated well with walnut mixture).

Bake for 50 minutes. In the meantime, prepare the syrup.

Walnut Cake (con't)

Insert a knife in the middle of the cake. Cake is done if the knife comes out clear. When cooked, turn the oven off and leave the cake in the oven for an additional 10 minutes.

Remove from oven and immediately pour cool syrup over the cake, one tablespoon at a time. Set cake aside to cool.

Cut into pieces and serve.

To Make the Syrup
Combine water with sugar; add cinnamon stick and simmer for 20 minutes. Remove from heat and discard cinnamon stick; stir in lemon juice and set aside to cool.

Tip: To decorate the cake, cut your designs on a piece of paper, place over cake and sprinkle with confectioners' sugar.

Preparation time: 20 min.   Baking time: 1 hr.   Yield: 18 pieces

Butter Cookies Covered with Confectioners' Sugar (*Kourabiedes*)

Kourabiedes are one of the primary sweets served on Christmas Day. It is traditional to serve this delight with brandy. Use the best, unsalted butter you can find to make these cookies. You will not be able to resist second helpings.

    12 ounces all-purpose flour
    1 teaspoon baking powder
    8 ounces unsalted butter (room temperature)
    3 ounces confectioners' sugar
    1 egg yolk
    3 tablespoons brandy
    1 cup blanched almonds, toasted, coarsely chopped
    3 tablespoons rose water

Decoration
    1 pound confectioners' sugar

In a medium mixing bowl, combine flour and baking powder. Sift twice; and set aside. In a mixer, beat butter until color turns almost white. Add confectioners' sugar and beat until sugar is well incorporated into butter. Add egg yolk, brandy, and mix well.

Make a well in the flour and add wet ingredients. At this point work dough by hand. Mix to achieve hard dough that does not stick to your fingers. Add almonds now and mix to incorporate. Do not overwork dough. Preheat oven to 350° F.

Place dough on a flat surface and roll out to about 3/4 inch thick. Shape into small oval pieces (about 2 x 1 x 3/4 inch) and arrange on baking sheets on top of parchment paper. You may make your own designs. Dough will not rise when cooked.

Bake on middle oven rack for 25–30 minutes or until Kourabiedes start to get a light golden color. They should not be browned. Remove from oven and immediately brush with rose water; set aside to cool. After 5 minutes and while still warm, sprinkle on a small amount of the confectioners' sugar and set aside to cool completely. Cover with the remaining confectioners' sugar to keep them fresh and soft.

Tip: When sprinkling confectioners' sugar, use a sifter to break up all the lumps and to keep the sugar fluffy. Do not handle Kourabiedes hot, as they may break.

Preparation time: 1 hr.   Baking time: 30 min.   Yield: 24–28 pieces

Honey-Dipped Cinnamon Cookies (Finikia Plagias)

This is the way we make Finikia in the town of Plagia, on the island of Mytilini, my birthplace. The cookies are dipped in syrup and decorated with ground walnuts and a cinnamon mixture.

    3 pounds all-purpose flour
    1 tablespoon baking powder
    8 ounces butter (room temperature)
    6 ounces Crisco (room temperature)
    1 1/2 cups sugar
    1 cup extra virgin olive oil
    3 tablespoons brandy
    3 tablespoons orange zest
    1/2-teaspoon baking soda
    1 cup orange juice
    4 eggs

Filling (optional)
    2 eggs
    1 cup walnuts, finely chopped
    1 teaspoon ground cinnamon

Syrup
    4 cups water
    2 cups sugar
    2 cups honey
    1 cinnamon stick

Decoration
    2 cups walnuts, finely chopped
    1 teaspoon ground cinnamon

In a large mixing bowl, mix flour with baking powder and sift twice. Set aside.
In a mixer beat butter, Crisco, and sugar until smooth. Add oil, brandy, and orange zest; mix on slow speed to incorporate the ingredients.

Dissolve baking soda into juice; add into the mixer bowl and mix for a few seconds. Add the eggs and mix to incorporate. Set aside. You can do this step by hand using a whisk.

Make a hole in the middle of the flour and pour in wet ingredients. With your hands, mix well to form a hard dough. Dough should not stick to your fingers.

Preheat oven to 350° F.

Place dough onto a flat surface and roll to about 1/2 inch thick. Cut dough into different shapes with cookie cutters and place on baking sheet covered with parchment paper. Space them about 1 inch apart.

Bake for 35–40 minutes or until golden. Remove from oven and set aside to cool overnight before you blanch in syrup.

To Make Filling and Stuff Finikia (optional)

Mix eggs, walnuts and cinnamon to make a soft paste.
To stuff Finikia, as in step 6 above, instead of rolling dough, make small balls the size of a walnut and then flatten them out to make patties (about 3 inches diameter).

Place 1/2 teaspoon filling on the dough patty and roll patty to enclose the filling. Place finikia on baking sheet with seam at the bottom. Bake as shown in step 7 above.

To Make the Syrup and Decorate Finikia:

In a medium saucepan, bring water to a boil. Add sugar, honey, and cinnamon stick; simmer for 10 minutes and then remove and discard cinnamon stick.

In a small mixing bowl, combine walnuts and ground cinnamon. Set aside.

Dip 3–4 Finikia at a time into boiling syrup for 30–40 seconds. Using large perforated spoon, remove Finikia and place onto a large dish. Do not leave Finikia in the syrup longer than 30–40 seconds, as they will become soft and will break. Continue dipping until all Finikia have been dipped into the syrup.

When Finikia are placed on the dish, sprinkle with walnuts. This way the walnuts will stick on the top.

Honey-Dipped Cookies (Con't)

Tip: The most prominent flavor in these cookies is the olive oil. Therefore use the best olive oil you can find. Because these cookies are associated with the celebration of Christmas, it is a tradition to give a festive shape to the cookies, so shape them like stars or pine trees.

Preparation time: 1 hr. 30 min.   Baking time: 40 min.   Yield: 6–7 dozen Finikia

Honey-Dipped Cookies.

Farina Cake
(Halvas me Simigdali)

The combination of roasted farina, butter, and syrup makes this an unusual, but very tasty and easy to make cake. You might need a little help to make this cake.

  8 ounces butter
  2 tablespoons pine nuts
  2 cups fine farina
  1 teaspoon ground cinnamon

Syrup
  1 cup sugar
  4 cups water
  1 cinnamon stick
  10 whole cloves

For this cake, you will make the syrup first. In a medium saucepan, combine sugar and water; add cinnamon stick, cloves, and bring to a boil; boil for 10 minutes. After 10 minutes, remove cinnamon and cloves and discard. Lower the heat and keep the syrup warm (not boiling) while preparing the rest of the ingredients.

In a medium saucepan, melt 2 ounces of the butter and roast pine nuts until light brown; then add remaining butter to melt; then add farina. Mix well and roast until farina is light brown, stirring continuously over medium heat. Pour hot syrup into saucepan with the farina, at once; this is the step when you will need someone to help you. Mix well. Reduce heat to very low and cook for 3–4   minutes stirring frequently, cover and simmer over very low heat. After 10 minutes, mix with a fork to aerate the mixture. Cover and simmer for another 10 minutes at very low heat.

Remove from heat, cover with paper towel, place a lid on it, and set aside to cool. Before serving use a wooden fork to break any lumps, aerate, and make the Halva fluffy. Sprinkle with cinnamon and serve cold.

Tip: If you follow the above instructions closely your Halva will be fluffy and will not lump. If your Halva is lumpy, then place into a greased mold, press Halva into the mold and let it cool. Remove from the mold, cut into serving-size pieces, sprinkle with cinnamon and serve. If you want your Halva to be white, then remove roasted pine nuts from butter and wipe saucepan clean before melting the remaining butter. Let pine nuts brown to give a nice contrast with the white farina. Also, do not let farina roast. Just mix with butter and go to next step.

Preparation time: 25 min.   Cooking time: 30 min.   Yield: 6–8 servings

Farina Cream Cake
(Bougatsa)

Desserts with custard are always a Greek favorite. This recipe is from my local baker and has not changed for the last 80 years. This was my treat every Sunday after church.

> 1 pound filo dough #4
> 20 ounces butter
> 1 cup fine farina
> 3 1/2 cups milk, hot
> 1 cup sugar
> 1 teaspoon vanilla extract
> 4 eggs, lightly beaten
> 1/2 cup confectioners' sugar
> 1 tablespoon ground cinnamon

Remove filo dough from refrigerator and leave at room temperature for 6 hours. If frozen, thaw at room temperature for at least 12 hours or overnight. Keep in original packaging while thawing.

In a medium saucepan, melt 4 ounces butter and sauté farina for 5 minutes.

Bring milk to a boil and pour slowly into farina mixture, mixing continually. Add sugar and vanilla extract, mixing continually until mixture thickens. You will make a thick cream. Remove from heat and set aside to cool for 10 minutes. Cream must be cool before you add the eggs.

Mix eggs with cream mixture. Melt remaining butter and grease a 17 x 12 x 1-inch baking pan.
Preheat oven to 350° F.

Unroll filo sheets and place one sheet flat in the baking pan. Brush with butter. Repeat with another 10–12 sheets, brushing each sheet with butter. Now spread the cream evenly over the filo sheets and cover with the remaining filo sheets one at a time, brushing each sheet with butter.

Pour remaining butter on top and set aside for 10 minutes. Trim excess filo sheets. Bake Bougatsa on the middle oven rack for about 45–50 minutes or until filo dough becomes golden.

Remove from oven and while hot, use a pizza cutter to cut into bite-sized pieces.

Sprinkle with confectioners' sugar and cinnamon. Serve hot.

Preparation time: 30 min.   Cooking time: 50 min.   Yield: 8–10 servings

Farina Cake.

Farina Cake with Syrup
(Ravani)

This is a sweet and delicious cake. It is excellent served with vanilla ice cream.

- 1 cup fine farina
- 1 tablespoon baking powder
- 3 tablespoons, all-purpose flour
- 9 eggs (separate yolks and whites)
- 1 cup sugar
- 1 teaspoon lemon zest
- 4 tablespoons brandy
- 24 almonds, blanched

Syrup
- 2 1/2 cups water
- 2 cups sugar
- 2 pieces of lemon rind
- 2 tablespoons lemon juice
- 1/2 stick (2 ounces) butter

Grease a 12 x 9 x 2-inch baking pan and set aside. Preheat oven to 350° F.

In a medium bowl, mix farina, baking powder, and flour; sift together once.

Beat egg whites into a soft meringue and set aside. You can beat them by hand or use an electric mixer.

Beat egg yolks with sugar until they obtain a pale yellow color and double in volume. Add lemon zest, brandy, and incorporate with egg yolks.

Slowly add farina mixture and mix until all ingredients are incorporated.

Lastly, fold in the meringue and keep the mixture fluffy. Ensure that meringue is well incorporated with the rest of the ingredients.

Pour mixture into baking pan, spreading across evenly.

In the meantime, make the syrup. The best time to finish the syrup is when you take the Ravani out of the oven.

Bake on middle oven rack for 45–50 minutes. Insert a knife into the middle of the cake. Ravani is done when knife comes out clear. Turn heat off and keep Ravani in the oven for an additional 10 minutes.

Remove from oven and immediately pour hot syrup over hot Ravani one tablespoon at a time. Cover with kitchen towel and set aside to cool for a couple of hours.
When cold, cut into small pieces, decorate each piece with almonds, and serve.

To Make the Syrup
In a medium saucepan, bring water to a boil. Add sugar and lemon rind; reduce heat and simmer for 15 minutes. Add lemon juice, remove from heat, and discard lemon rind. Add butter and mix until butter is melted.

Preparation time: 30 min.   Cooking time: 1 hr.   Yield: 20–24 pieces

Butter Cookies
(*Koulourakia Marikas*)

My sister makes these cookies and they are delicious. They make the perfect accompaniment for your morning coffee or afternoon tea.

> 1 1/4 pounds all-purpose flour
> 4 eggs (separate yolks and whites)
> 1 1/2 cups (12 ounces) sugar
> 4 ounces butter (room temperature)
> 1/2 teaspoon ammonia powder
> 1 teaspoon vanilla extract
> 2 tablespoons orange zest
> Decoration
> 1 egg
> 1 tablespoon white sesame seeds
> 1/4 cup milk, lukewarm

In a medium mixing bowl, sift flour twice and set aside.

In a mixer bowl, beat yolks with half the sugar until they turn a pale yellow color and mixture is fluffy and has doubled in size.

In another bowl, beat butter with remaining sugar and then combine with yolk mixture. Beat egg whites to form a soft meringue; fold into egg yolk mixture.

Dissolve ammonia in milk, add vanilla extract, and pour into egg yolk mixture. Add orange zest. Slowly add flour into egg yolk mixture, mixing slowly to make a soft dough. Cover dough with plastic wrap and refrigerate for 15 minutes.

Preheat oven to 350° F.

Grease a large baking sheet. Roll small pieces of dough (the size of a large walnut) into 6-inch-long sausage and then form into the desired shape (like the number 8 or the letter Y). Arrange cookies on a baking sheet spacing them at least 1 inch apart. These cookies will rise and will double in thickness when baked.

To decorate the cookies, beat egg; brush the cookies and sprinkle with some sesame seeds. Bake on middle oven rack for 25-30 minutes or until a light golden color.

Tip: If the dough sticks when you roll it, use very small amounts of flour to just dust the surface. Do not add more flour to the dough.

Preparation time: 1 hr. 20 min.   Cooking time: 30 min.   Yield: 4–5 dozen cookies

Butter Cookies.

Honey-Dipped Doughnuts
(Loukoumades)

This is the authentic Greek doughnut. On the island of Mytilini in the olden days, the farmers celebrated the end of the olive picking season with Loukoumades.

    8 ounces all-purpose flour
    1 ounce fast-acting dry yeast
    1/4 cup lukewarm water (110–115° F)
    2 ounces butter, melted
    1/2 teaspoon salt, dissolved in 1 cup lukewarm water
    4 cups olive oil

Syrup
    2 cups sugar
    1 cup water
    1/4 cup honey
    2 tablespoons lemon juice
    1/4 cup walnuts, finely chopped
    1/2 teaspoon ground cinnamon

In a medium plastic bowl, dissolve yeast in 1/4 cup water and add 1 tablespoon flour. With a fork, mix well until smooth. Cover with plastic wrap and kitchen towel and set aside in a warm place for about 30 minutes to rise. Yeast should start foaming in 4–5 minutes. If not, replace yeast.

In the meantime prepare the syrup. In a medium saucepan, combine sugar, water, honey, and bring to a boil. After 5 minutes over high heat, reduce heat, and skim off and discard foam. Simmer for an additional 10 minutes. Add lemon juice, remove from heat, and set aside to cool.

In a large plastic mixing bowl, add remaining flour and make a well in the middle. Add yeast, butter, and slowly add water with salt, mixing to make soft, smooth, but elastic dough. Dough should be soft so that it spreads in the bowl by itself. The consistency of the dough is very important in order to make golden and crispy Loukoumades.

Cover bowl with plastic wrap and kitchen towel and set aside in a warm place to rise for 2–3 hours. Punch dough down, work it for a couple of minutes, and test its elasticity.

Dough is ready when you can pull it for 6–8 or more inches without breaking it.

In a medium-heavy saucepan, heat oil to about 375° F.

Take some of the dough in your hand and squeeze it slowly so that a dollop of dough the size of a small walnut emerges up between your thumb and index finger. Remove dollop of dough with a wet spoon and slowly place into hot oil. If oil is hot enough, dough will drop to the bottom of the saucepan, but will pop up to the surface in a few seconds. Continue wetting the spoon in cold water as you make Loukoumades. This way the dough will not stick to the spoon. Keep your heat high at all times while frying Loukoumades. Try to keep Loukoumades emerged in the hot oil so that they cook on all sides at the same time.

When Loukoumades are golden, remove from the oil with a large perforated spoon and place directly into cold syrup. Press down on top to cover Loukoumades with syrup. Keep Loukoumades in the syrup for 5 minutes.

While still hot, place a few Loukoumades into a serving bowl, pour some honey over the doughnut, sprinkle it with walnuts and cinnamon, and serve.

Preparation time: 20 min.   Cooking time: 30 min.   Yield: 2–3 dozen bite-sized
doughnuts

Honey-Dipped Doughnuts.

Cake with Apricot Preserves
(Pasta Flora)

The dough is buttery and rich, but the introduction of the apricot preserves on top of the cake gives it a very nice color and flavor.

  3 cups (12 ounces) white flour
  1 cup sugar
  8 ounces butter, room temperature
  1 tablespoon baking powder
  3 egg yolks
  3 tablespoons brandy
  1 teaspoon orange zest
  1 cup apricot preserves
  3 tablespoons milk

In a medium mixing bowl combine sugar and butter and beat until smooth.

In another bowl combine flour with baking powder and then mix egg yolks, brandy, and orange zest; now combine with sugar and butter.

Wrap in plastic foil and refrigerate for 30 minutes.

Reserve about 1/2 cup of the dough for decoration. Place the remaining dough in an oiled 8-inch pie dish or 8 x 2-inch-square baking pan and spread evenly. Spread apricot preserves evenly over dough. Preheat oven to 350° F.

Mix milk with reserved dough, add 1 tablespoon water and roll out into a thin sausage. Cut into long strips. Arrange the strips on top of preserves in a crisscross pattern to form a lattice.

Cover with aluminum foil and bake on middle oven rack for 25 minutes. Then remove foil and bake for an additional 20 minutes.

Remove from oven, cut into pieces, and serve cold.

Preparation time: 30 min.   Cooking time: 45 min.   Yield: 9–18 pieces

Butter Cookies with Mastich
(*Koulouria Mastihas*)

These cookies have the very distinctive flavor of the Mastich. This aromatic ingredient is sold dry (small pieces or powder) and it is produced in one place in the world, on the Aegean island of Chios.

    1 pound flour
    1 pound butter, room temperature
    2 cups sugar
    4 eggs
    3/4 cup heavy cream
    1 teaspoon cinnamon
    1/2 teaspoon Mastich Chiou
    1 teaspoon vanilla extract
    6 teaspoons baking powder

Decoration
    1 egg yolk, dissolved in 1 tablespoon milk
    1 tablespoon white sesame seeds

In a mixing bowl, beat butter and sugar until mixture is smooth. Add eggs one at a time and beat well. Then add cream and mix for a couple of minutes.

Add cinnamon, mastich, vanilla, and beat.

Mix baking powder and flour; sift twice. Slowly combine the dry ingredients with the wet to fully incorporate them. Work the dough until it is soft, but not sticky. Let dough rest for 10 minutes. Preheat oven to 350° F.

In the meantime, grease a large baking sheet. On a flat surface, roll out dough and make several different shapes of cookies with cookie cutters. Arrange cookies on a baking sheet covered with parchment paper.

Brush with egg yolk mixture and sprinkle with sesame seeds. Bake for about 25–30 minutes or until golden.

Remove and set aside to cool before removing from baking sheet.

Preparation time: 1 hr. 20 min.   Cooking time: 30 min.   Yield: 4 dozen cookies

Ring-Shaped Cookies
(Halkadakia)

These cookies are very simple to make. The main ingredient is the extra virgin olive oil. You will taste the aroma and the freshness of the oil. Use the best oil you can find.

- 1 pound all-purpose flour
- 1 tablespoon baking powder
- 1 cup extra virgin olive oil
- 2 tablespoons sugar
- 1 cup white wine
- 1 teaspoon salt

Combine flour with baking powder, sift twice, and set aside.

Mix oil and sugar until smooth.

Add wine, salt, and mix well. Slowly add flour and mix to form a soft dough. Do not overwork dough. Let dough rest for 10 minutes. Preheat oven to 350° F.

In the meantime, grease a large baking sheet.

Using dough pieces the size of a small egg, roll to make a sausage-like shaped dough about 6 inches long. Make ring-shaped cookies and arrange on baking sheet.

Bake on middle oven rack for 15–20 minutes. The cookies are done when the bottom is golden brown. Remove from oven and set aside to cool.

You can preserve the cookies by placing them in a cookie jar. Seal well and cookies will stay fresh 2–3 weeks.

Preparation time: 1 hr.   Cooking time: 20 min.   Yield: 3–4 dozen cookies

Crispy Cookies
(Paximadia Glyka)

These crispy cookies are great anytime, but they are fantastic with coffee. Store in a cookie jar and enjoy them for a few weeks.

   2 pounds all purpose flour
   5 tablespoons baking powder
   1 pound butter, room temperature
   3 cups sugar
   10 eggs
   1 tablespoon vanilla extract
   1/2 teaspoon ground cloves
   2 cups blanched almonds, roasted and sliced
   2 egg yolks
   1 tablespoon white sesame seeds

Mix baking powder with flour, sift twice, and set aside.

In a mixer, beat butter and sugar until soft and fluffy. Add eggs one at a time, beating continually until they double in volume. Add vanilla and cloves; mix.

Slowly incorporate flour into egg mixture to make a soft dough. Add almonds and mix again to incorporate with the rest of the ingredients. Preheat oven to 350° F. In the meantime, grease a baking dish.

Shape dough into two loaves (like a baguette) and place on baking dish leaving 3–5 inches between the loaves. Brush with egg yolks and sprinkle with sesame seeds. Bake for 20 minutes.

Remove from oven and set aside to cool. When cool, slice loaves to about 3/4 inches thick and arrange flat on a baking sheet. Loaves must be cool before you slice them.

Lower temperature to 225° F and bake for an additional 40 minutes. Remove from oven, turn cookies over, and bake for an additional 40 minutes. At this temperature you want to dry the cookies. Remove and set aside to cool.

Preparation time: 1 hr.   Cooking time: 1 hr. 40 min.   Yield: 3 dozen cookies

Rice Pudding
(Rizogalo)

This is a very popular pudding in the Greek cuisine. In small towns, rice pudding is made with goat's milk. It is very tasty.

    5 cups water
    1/3 cup rice
    1/2 teaspoon salt
    7 cups milk
    Lemon peel
    1 tablespoon corn flour, dissolved in 4 tablespoons milk
    3/4 cup sugar
    2 tablespoons white raisins
    1 tablespoon ground cinnamon

In a large saucepan, bring water to a boil. Add rice and salt. Reduce heat to very low and simmer, uncovered, until rice is soft and the water is fully evaporated.

Add milk and lemon peel; simmer at low heat, stirring frequently. After 20 minutes add dissolved cornstarch and stir constantly for 10 minutes.

Add 3/4-cup sugar, raisins, and simmer for an additional 10 minutes, stirring constantly.

Remove from heat and pour into serving bowls. Let it cool and then refrigerate. When ready to serve, sprinkle with cinnamon.

An alternate way of serving this dish is to sprinkle 2 tablespoons of sugar over the serving bowls with the rice pudding; then use a torch to burn the sugar to caramelize the surface. Sprinkle with cinnamon and serve.

Preparation time: 10 min.   Cooking time: 2 hrs.   Yield: 8 servings

Yogurt
(Yiaourti)

Greeks eat a lot of yogurt. They eat it plain after the main meal or they make Tzatziki sauce. It is very popular to eat yogurt covered with honey and sprinkled with walnuts.

    1 quart milk
    1 pint light cream
    5 tablespoons plain yogurt (this will be the starter)

In a medium saucepan, bring milk and light cream to a boil. Reduce the heat and simmer, stirring frequently so that the mixture does not stick to the bottom of the saucepan. Simmer for 30 minutes.

Remove from heat and pour into 1, 2, or 3, heavy ceramic or Pyrex bowls; leaving about 1/2 inch at the top to allow room to add the yogurt. In a small mixing bowl, reserve 1/4 cup of the milk.

Place bowls of warm milk on top of a kitchen towel and let cool down to about 120-130° F, or until you can hold your little finger in the milk for at least 10 seconds.

Dissolve 5 tablespoons yogurt into the bowl of reserved milk. Stir equal parts of the mixture into the bowl(s) of lukewarm milk. If you would like to keep the skin formed on top of the milk intact, then add dissolved yogurt in one end of the bowl(s) and slowly stir underneath the skin.

Place a couple of skewers or utensils on top of each bowl and cover well with kitchen towels. This way the towels will not touch and damage the top of the yogurt. Leave for 8 hours to set and then refrigerate.

Tip: If you prefer to serve a thicker yogurt, after yogurt has cooled in the refrigerator, empty into cheesecloth and suspend to drain for 2 hours. This way the excess liquid will drain out.

Preparation time: 20 min.   Cooking time: 30 min.   Yield: 4 servings

Babas
(Babadakia me Brandy)

This is a hearty sweet, baked in the oven, bathed in syrup, and then basted with apricot preserves. It's excellent with ice cream.

> 1 ounce (4 envelopes) fast-acting yeast
> 1 cup milk, lukewarm (110–115° F)
> 2 1/2-cups (8 ounces) self-rising flour
> 4 eggs, beaten
> 4 ounces (1 stick) butter, melted
> 1/8 teaspoon salt
> 2 tablespoons sugar
> 1/2 cup currants

Syrup
> 1/2 cup honey
> 1 1/2 cup water
> 1 cup sugar
> 1 piece of lemon rind
> 4 tablespoons brandy

Decoration
> 1 cup apricot preserve
> 3 tablespoons water
> 1/2-cup pistachio nuts, chopped

In a medium-size mixing bowl, dissolve yeast in the milk. Add 2 tablespoons flour and stir to make a smooth mixture. Set aside to rise. It is ready when frothy; about 20-25 minutes.

In the same bowl, add remaining flour, eggs, butter, salt, and sugar; beat for 4-5 minutes. Add currants and mix to blend. Set aside for 30 minutes or until batter doubles in volume.Grease 12 (6 ounces) custard cups.

Punch down the mixture, stir a couple times with a spoon, and pour mixture into custard cups. Fill cups half way to allow room for the mixture to rise. Set aside to rise until cup is three-quarters full. This will take about 10–15 minutes. In the meantime, preheat oven to 375° F.

Bake Babas on the middle oven rack for 20–25 minutes.

In the meantime, make the syrup (see directions below).

When Babas are ready, remove from the oven and pour syrup over each custard cup, one tablespoon at a time. Ensure that the Babas absorb the syrup. If the syrup is running off the pastry, make a few holes in the Babas using a toothpick.

In a small mixing bowl, combine apricot preserves with the water and stir to make a smooth paste. Remove Babas from custard cups, baste with the preserves, and sprinkle with the pistachio nuts.

To Make Syrup
In a medium saucepan, combine the honey, water, sugar, and lemon rind; bring to a boil. Reduce the heat and simmer for 5 minutes. Remove from heat and add the brandy. Stir to blend. Discard lemon rind.

Preparation time: 1 hr. 30 min.   Cooking time: 25 min.   Yield: 10-14 Babas

Honey-Covered Pastry
(Diples)

This dessert is easy to make. You can make it into different shapes while it is being fried. Cover with honey and walnuts, and serve hot or cold.

>2 cups self-rising flour
>4 eggs, beaten with 2 tablespoons sugar
>2 tablespoons virgin olive oil
>1 teaspoon lemon zest

For Frying
>Flour for dusting
>2–3 cups olive oil

Syrup
>2 cups honey
>1 cup water

Decoration
>1 tablespoon ground cinnamon
>1 cup walnuts, finely chopped

In a mixing bowl, add 1 1/2 cups of the flour and make a well. Add eggs, oil, and lemon zest in the well and mix together until you make a smooth, but hard dough. Add more of the remaining flour, if required, to achieve the right consistency.

Shape dough into a ball, cover with plastic wrap, and set aside for 1 hour.

Cut dough into small pieces and roll out onto a flat surface in very thin sheets. Sheets should be about 4 x 4-inch squares. Dust dough with flour if dough is sticking.

In a deep frying pan or a large saucepan, heat oil. When oil is hot, fold the dough pieces to form a cylinder and fry in the oil. Fry both sides until golden. Remove from heat and place on paper towel to drain any excess oil.

While pastries are draining, prepare the syrup (directions below).

Dip fried pastry into syrup and let it simmer for 5–10 seconds and remove. Place on serving dish and sprinkle with cinnamon and walnuts.

To Make the Syrup

In a medium saucepan, combine, honey, water, and bring to a boil. Reduce heat and simmer for 5 minutes. Remove foam and discard.

Preparation time: 30 min.   Cooking time: 30 min.   Yield: 30–36 pieces

TRADITIONAL SWEETS

Greeks love their traditional sweets, especially in the afternoon after their   summer siesta. Greek sweets are made out of fresh fruit and are preserved in jars year-round. They are served on a small dessert plate accompanied by ice-cold water and an aromatic, freshly ground, Greek coffee.

Greeks prefer to eat fresh fruit after their lunch or dinner. From fresh sweet cherries, juicy watermelon, or aromatic pears and grapes in the summer, to beautiful apples and blood-red oranges in the winter, these make up the majority of the fresh fruits. Almost every fruit, which is eaten fresh, is processed as a preserve, a jelly, or as whole fruit. When you visit Greece, you must try some of these sweets. It is an unforgettable experience.

In the last 20 or 30 years, the only sweets available were those made by mass production. Since late 1990, on the island of Mytilini, groups of women have started cooperatives to produce traditional homemade sweets. When you visit Greece, be sure to purchase some. They are great and stay fresh for a long time.

All locally grown fruit like citrus, pears, peaches, apricots, cherries, figs, pistachios, grapes, walnuts, tomatoes, and even eggplants—are prepared and cooked in syrup. Fruits need to be ripe but firm, and free of spots.

Apricot Preserves
(Verikoka Glyco)

Local apricots are available in Greece from early June to late July. They are sweet and excellent served as fresh fruit, dried fruit, or preserves.

    2 pounds apricots (ripe but firm)
    1/2-cup baking soda
    4 cups water

Syrup
    3 cups water
    2 pounds sugar
    1 cinnamon stick
    4 tablespoons lemon juice

In a large mixing bowl, dissolve baking soda in water and set aside.

With a sharp knife, peel apricots. Remove pits without damaging the fruit. Place them into the soda water. After soaking in the soda for 3 hours, rinse the apricots 3–4 times with cold water and drain.

To make the syrup, combine water with sugar and lemon juice in a medium saucepan. Add a cinnamon stick and bring to a boil for 5 minutes. Skim foam and discard. Add apricots and simmer for 15 minutes. Do not cover saucepan.

Remove apricots and set aside in a jar and continue boiling the syrup for another 10–15 minutes, or until syrup becomes thick. One way to test the thickness of the syrup is to pour a couple of drops of the hot syrup into ice water. When the syrup drop stays intact and on the surface, the syrup is ready.

Remove syrup from heat and set aside to cool. Discard cinnamon stick. When cool pour into jar with apricots, cover and store. You can keep the preserves for months.

Preparation time: 30 min.   Cooking time: 30-40 min.   Yield: 6 cups

Sour Cherry Preserves
(Vissino Glyko)

This is a very popular preserve, if not the most popular. It is served alone or with vanilla ice cream. Syrup is mixed with cold water and it is served as a vissinada drink, a very refreshing summer drink.

    2 pounds sour cherries (ripe but firm)
    1 pound sugar
    1/4 cup water
    4 tablespoons lemon juice or 1 teaspoon citric acid powder
    2–3 rose-scented geranium leaves (optional)

Wash cherries and remove pits. Try to avoid breaking the fruit. Preserve juices.

Place cherries into a medium saucepan in layers and sprinkle each layer with sugar. Add juices and water; set aside for few hours or overnight.

Bring cherry mixture to a boil; reduce heat and simmer for 15 minutes. Remove cherries with a perforated spoon and set aside.

Add the geranium leaves and continue boiling the syrup for another 20–25 minutes, or until mixture becomes the consistency of loose honey. Stir gently and frequently while simmering.

Add lemon juice or citric acid, stir, and remove from heat. When cool, place sour cherries into a jar, pour the syrup over to cover the cherries, cover with a lid, and preserve.

Tip: Preserve excess syrup. Use it over vanilla ice cream, or make vissinada. To make vissinada, dissolve two tablespoons of the syrup in a glass of ice water and offer it as a cool drink. This is excellent on a hot, summer afternoon after siesta. Before you pour the syrup over the cherries, let the syrup cool. This way you will be able to tell if it needs more thickening.

Preparation time: 60 min.   Cooking time: 40 min.   Yield: 4 cups

Cherry Preserves
(Kerasi Glyko)

This is another popular preserve in Greece. This is used as a sweet in the afternoon, after the siesta. It is similar in texture to sour cherry preserves, but does not have the sweet and sour flavor.

    2 pounds cherries (ripe but firm)
    12 ounces sugar
    1 teaspoon vanilla extract
    10 whole cloves (in spice bag)
    4 tablespoons lemon juice or 1 teaspoon citric acid powder

Wash cherries and remove pits. Try not to damage the fruit. Preserve juices.

Place cherries into a medium saucepan in layers and sprinkle each layer with sugar. Add juices, vanilla, and cloves; set aside for a few hours or overnight.

Bring cherry mixture to a boil and simmer for 15 minutes. Remove cherries with a perforated spoon and set aside. Continue boiling the syrup for another 20–25 minutes or until syrup thickens and becomes the consistency of loose honey. Stir gently and frequently.

Add lemon juice or citric acid, mix, and remove from heat. Remove and discard cloves.

When cool, place cherries into a jar and pour syrup over the cherries, cover with a lid, and preserve.

Tip: Before you pour syrup over the cherries, let the syrup cool. This way you will be able to tell if it needs more thickening.

Preparation time: 60 min.   Cooking time: 40 min.   Yield: 4 cups

Orange Peel Preserves
(Portokali Glyko)

This is a fantastic preserve. It is very colorful and has a tangy flavor.

- 4 large oranges with thick skins (to make about 1 pound of orange peels)
- 1 1/2 pounds sugar
- 2 1/2 cups water
- 4 tablespoons lemon juice
- Needle and cotton string

Wash oranges and scrape lightly to remove wax.

Remove peel in the shape of long triangles. Roll to form cylindrical shapes with triangle base on the inside of the roll. Use a needle and a string to pull through the rolls (like a necklace) to maintain their shape until cooked.

In a medium saucepan, add enough water to cover the orange peel and bring to a boil. Boil for 10 minutes. Drain and replace water and boil again. Repeat this step three times, and then drain, remove orange peel, and set aside to cool.

In the same saucepan, add sugar, 2 1/2 cups water and boil for 5 minutes. Add oranges and boil an additional 40 minutes.

Add lemon juice, stir, and remove from heat; set aside to cool. When cool to the touch, cut and remove the string. Place orange pieces into a jar, cover with syrup and a lid, and preserve.

Tip: Oranges should have very thick skins. Use a cheese grater to scrape the skin of the oranges to remove any wax.

Preparation time: 30 min.   Cooking time: 1 hr. 20 min.   Yield: 4 cups

Quince Preserves
(*Kidoni Glyko*)

A popular preserve made from ripe quince. Very aromatic and frequently used as a spread on toasted bread.

> 2 pounds quinces, ripe
> 2 lemons, juice of
> 1 1/2 cups quince stock (see below)
> 3 1/2 cups sugar
> 10 whole cloves
> 3-4 rose-scented geranium leaves (optional)
> 1/2-teaspoon rose water

In a medium bowl, combine 5 cups water and lemon juice from one of the lemons; set aside. In a medium saucepan, bring 5 cups water to a boil.

In the meantime, peel quince and remove seeds. Place seeds, peel, and flesh all in the boiling water; cover and simmer for 15 minutes.

After 15 minutes, remove flesh and cool in ice water; continue simmering the seeds and the peel for an additional 15 minutes. Remove saucepan from heat, strain and preserve 1 1/2 cups of the stock; discard peels and seeds.

Coarsely grate the flesh of the quince and place into the bowl of lemon water to prevent discoloring.

Return preserved stock to the saucepan and bring to a boil. Add sugar, cloves, and geranium leaves, if using.

Add grated quince and simmer for about 50 minutes or until syrup thickens. Mix frequently.

Remove from heat and stir in lemon juice from the remaining lemon and rose water. Cool and pour into a jar. Close lid and preserve.

Tip: Quince is available in Greece from October to December. In the U.S. they are in bloom in the Spring.

Preparation time: 30 min.   Cooking time: 1 hr. 30 min.   Yield: 4 cups

Quince Paste
(Kidonopasto)

This fruit is so fragrant that if placed in a room, it will fill it with a strong aroma.

> 3 pounds quince, ripe
> 3–4 rose-scented geranium leaves
> 1 cup quince stock (see below)
> 2 cups sugar (about 1 cup of sugar per cup of quince puree)
> 2 tablespoons lemon juice
> 1/2 teaspoon rose water
> 20-24 cloves, whole
> 1/2-cup sugar to cover Kidonopasto pieces

In a medium saucepan, bring 5 cups water to a boil.

Slice quince into quarters, peel, and remove the seeds. Place peels and seeds in cheesecloth, tie, and put in boiling water. Add flesh pieces and geranium leaves, cover, and simmer until flesh is soft.

Remove from heat, drain, and preserve flesh and 1 cup stock. Discard remaining ingredients. Puree soft quince.

In the same saucepan, bring to a boil preserved quince stock, add quince puree and 1 cup of sugar per cup of quince puree. Add lemon juice. Simmer, stirring constantly until mixture comes away from the side of the pan.

When thickened, add rose water and remove from heat; pour into a 12 x 9 x 1-inch baking pan covered with parchment paper.

Set aside to cool and cut into different shapes like squares, diamonds, or rounds. Place 1 clove on top of each piece and set aside, uncovered, to dry for a day or two. Roll each piece in sugar and store.

Preparation time: 40 min.   Cooking time: 1 hr. 20 min.   Yield: 18–24 pieces

Sour Cherry Liquor
(Brandy me Vissino)

This liquor is made in July using fresh sour cherries and aromatic spices. Although it will keep for long time, locals make a new batch of liquor every year. This is great with Kourabiedes (butter cookies) or Finikia (cinnamon cookies).

1/2 pound sour cherries
750 ml brandy
1/4 cup sugar
1 cinnamon stick
10 whole cloves

Wash cherries, remove stems, and discard.

Place all ingredients into a large jar, cover with plastic wrap, and set in a sunny place for 4 weeks.

Using a coffee filter, strain the brandy and pour into a bottle. Discard cherries and all other ingredients.

Preserves for a long time.

Preparation time: 20 min.   Yield: 1 1/2 pints

Turkish Delight
(Loukoumia)

This is a bite-size sweet served with coffee after dinner. It is covered with powdered sugar and is a sweet, delicious treat.

1/2 cup cornstarch
2 cups (6 ounces) confectioners' sugar
2 cups water
2 1/2 cups (1 pound) sugar
1/8 teaspoon citric acid powder or 4 tablespoons lemon juice
1 teaspoon ground Mastich Chiou
8 ounces confectioners' sugar for decoration
3/4 teaspoon vanilla extract
1 tablespoon rose water, or a flavor of your choice
2 drops of food coloring of your choice
1/2 cup almonds, blanched, toasted, chopped

In a medium mixing bowl, dissolve cornstarch and 2 cups of confectioners' sugar in 1/2 cup of the water and set aside. Mix well to remove lumps.

In medium saucepan, pour remaining water, add sugar, and citric acid or lemon juice, and bring to a boil. Reduce heat and simmer for approximately 5 minutes. If you are using Mastich, add it now.

Add dissolved cornstarch mixture to the saucepan and bring to a boil. Stir to break up any lumps. Continue boiling for at least 30–35 minutes until the cornstarch becomes a pale yellow. Do not shorten the boiling time. After 25 minutes, the mixture will become a translucent color. Give the boiling mixture frequent, rigorous stirs. You will notice that the mixture is becoming thick and is forming large bubbles as it is boiling.

Dust the bottom of a small, 8 x 10 x 1-inch metal tray with confectioners' sugar and set aside.

When the mixture is ready, add vanilla extract, rose water, or any other flavor you are using. Pour half of the mixture into the metal tray and add the food coloring to the remaining mixture; stir well to incorporate the color. You can color the entire mixture, if you prefer.

Place almonds on top of the mixture that is already in the tray and pour the remaining mixture on top. Dust with sifted confectioners' sugar. Let cool for 6–8 hours.

Cut mixture into bite-sized pieces and toss in confectioners' sugar; cover with remaining sugar and store in a box for a few weeks.

Tip: When you make this sweet, you must take the time to boil the mixture as indicated above. If you cook it for less time, the Loukoumia will be soft and sticky. If they are overcooked, they become chewy. Also, you may substitute pistachio nuts or hazelnuts for the almonds.

Preparation time: 20 min.   Cooking time: 35–40 min.   Yield: 30–36 Loukoumia

BREADS

Greeks love their breads. Up until the last few years, when people became overly concerned about counting calories, bread was very popular served with all types of food.

In the summertime, crusty bread, dipped in the juices of tomato salad, makes a delightful meal. Another way to enjoy good crusty bread is to toast it over charcoal and then sprinkle it with olive oil and sea salt.

Many times Greeks will take warm bread, slice it, and make a simple, but very refreshing sandwich using only Feta cheese, fresh tomato slices, and a couple of basil leaves.

There are many types of bread and they vary from place to place. The bread recipes provided in this book are those that are very popular on the island of Mytilini.

Making bread takes time, as you must wait for the starter and the dough to rise. However, not all steps have to be done at the same time. Just plan your other activities in between the time it takes for the starter and the dough to rise, and your bread-making adventure will be very easy. There is very little as pleasing or exciting as having a warm piece of homemade bread served with your meal. Bake some homemade bread for your family and friends and you'll see what I mean.

Soft Pita Bread
(*Pita*)

This is a popular bread that's oh-so-easy to make.

    2 cups all-purpose, unbleached, white flour
    1 teaspoon salt
    1/4 cup extra virgin olive oil
    9–10 tablespoons lukewarm water

In a large bowl, mix salt and flour; sift together twice. Make a well in the center of the mixture, add 2 tablespoons of oil and begin mixing. While mixing, slowly add water, one tablespoon at a time, until it forms a hard dough.

Knead dough on a lightly-floured surface for about 5–7 minutes, until the dough is smooth and elastic. Place into a medium lightly-oiled non-metallic bowl, cover with oiled plastic wrap, and let it rest for 30 minutes.

Divide the dough into 5 equal pieces and roll each into 6-8-inch rounds.

Heat a heavy frying pan over a medium heat.

Lightly oil the hot frying pan with 1 tablespoon of oil and cook one pita at a time for about 2 minutes on each side, or until it starts to brown.

Oil the frying pan every time you add a new pita.

Serve warm or cold. Excellent with Taramasalata, eggplant dip, or Tzatziki sauce.

Preparation time: 40 min.   Baking time: 20 min.   Yield: 5 pitas

Pita Bread

This is the pita bread you will find in the supermarkets. I like this pita lightly toasted and then filled with Feta cheese and a few slices of fresh, ripe tomatoes.

  2 cups all-purpose flour
  1 teaspoon salt
  1/2-ounce (2 envelopes) active dry yeast
  1 cup lukewarm water (110–115° F)
  2 tablespoons extra virgin olive oil

Sift flour and salt together. In a medium bowl, mix the yeast with the water until dissolved, add oil, and mix well.

Gradually mix flour into the yeast mixture and knead to make a soft dough.

Knead the dough on a lightly floured surface for 5–7 minutes, until dough becomes smooth and elastic. Place into a medium non-metallic bowl, cover with oiled plastic wrap and a kitchen towel, and let rise in a warm place for 1–2 hours, or until it doubles in volume.

Punch the dough down and on a lightly floured surface, divide the dough into 5 equal pieces. Roll out each piece into an oval shape pita about 1/4 inch thick and 6 inches long.

Place on a flat surface on top of a kitchen towel, cover with oiled plastic wrap and another kitchen towel, and let rise for approximately 30 minutes.

Preheat oven to 450° F and heat a 15 x 12 x 1-inch baking sheet for at least 5 minutes. If you have a baking stone, it would be better. Make sure the baking pan or stone is hot.

Place one or two pitas on the baking sheet and bake for 5–6 minutes, or until puffed up.

Let pitas cool. Serve hot, cold, or toasted.

Preparation time: 20 min.   Baking time: 20 min.   Yield: 5 pitas

Country Style Bread
(Psomi Horiatiko)

This bread is made when corn meal is available. The bread is rich and heavy, and it is great when served as an accompaniment to any soup, but especially to bean soup.

Starter
> 1 ounce (4 envelopes) active dry yeast
> 1 cup lukewarm water (110–115° F)
> 1 cup flour
> 1/2-cup honey

Dough
> 1 pound white all-purpose flour
> 1 pound whole wheat flour
> 1 pound yellow corn meal
> 1 teaspoon salt
> 3 tablespoons virgin olive oil
> 2 tablespoons sugar
> 1/4-cup warm milk
> 4 cups lukewarm water (110–115° F)
> Additional flour for rolling out dough

To Make the Starter
In a medium, non-metallic bowl, dissolve yeast in water. Add flour, honey, and mix until dough is smooth and has the consistency of a thick batter (like pancake batter).

Cover with oiled plastic wrap and a kitchen towel, and leave to rise in a warm place. The starter will foam and rise to 2–3 times its volume, and then will fall back, which indicates that it is ready to use. This process normally takes 1–2 hours, depending on the room temperature.

To Make the Dough
Mix flour and corn meal with salt and sift into a large bowl. Make a well in the middle
of the flour mixture. Add starter, oil, sugar, milk, and slowly add water, mixing together to form soft, elastic dough.

Cover the bowl with a moist kitchen towel and let dough rise in a warm place. When doubled in volume (takes about 3–4 hours), punch down the dough and knead on a lightly-floured surface for about 10 minutes.

Country Style Bread (con't)

Divide dough into 3 pieces, shape into baguettes, and place on baking sheets; leave some space between the loaves.

Cover with kitchen towel and let rise for about 30–40 minutes.

Preheat oven to 450° F and place a bowl of boiling water on the bottom oven rack to create moisture.

With a very sharp knife, make 3–4 slits diagonally across the top of the loaves.

Bake on middle oven rack for 40–45 minutes.

Bread is ready when the top becomes a nice, light, golden color.

Remove from the oven, brush with water, and let cool.

Preparation time: 30 min.   Baking time: 45 min.   Yield: 3 baguettes

Country Style Bread.

Pita Bread with Oil
(Pita me Ladi)

This bread was popular with farmers, when they made their own bread. But, as many other things, it is not easy to find this bread, unless your have good friends who make their own bread.

> 1 pound white bread flour (preferably flour with high gluten)
> 1 1/2 teaspoons salt
> 1 ounce (4 envelopes) active dry yeast
> 2 1/4 cups lukewarm water (110–115° F)
> 2 tablespoons olive oil
> 2 tablespoons fresh oregano
> 1 tablespoon sesame seeds

Sift flour and salt together into a large bowl and make a well in the center and set aside. In a small, non-metallic mixing bowl, mix yeast with 4 tablespoons of lukewarm water until smooth. Stir in the remaining water and then stir in the oil.

Pour yeast mixture into the flour well and mix into a soft dough. On a lightly-floured surface, knead the dough for 10 minutes until smooth and elastic.

Place into a lightly-oiled bowl, cover with oiled plastic wrap and a kitchen towel, and let rise in a warm place, until it doubles in volume. This will take about 2 hours.

Punch the dough down and gently knead in fresh oregano.

Divide the dough into 2 or 3 equal pieces and shape each into a ball. Roll each ball out into a 10-inch round.

Lightly oil 2 or 3 round baking dishes and place a pita into each pan. Cover with oiled plastic wrap and a kitchen towel and let rise in a warm place for approximately 45 minutes. In the meantime, preheat oven to 400° F.

When dough is ready, uncover it, and using your fingertips, make deep dimples over the entire surface of each pita. Brush each pita with water, sprinkle with sesame seeds, and bake at 400° F for 30 minutes, or until golden.

Remove from oven and let cool. Best when served warm.

Preparation time: 30 min.   Baking time: 30 min.   Yield: 2–3 pitas

Crusty Bread
(Greek Style Bread)

This is a very popular Greek bread. It is especially delicious when served with hearty bean soup or with Greek tomato salad.

Starter
>   1/4 ounce (1 envelope) active dry yeast
>   1/2 teaspoon sugar
>   1/2 cup lukewarm water (110–115° F)
>   3/4 cup unbleached white flour

Dough
>   1 pound unbleached, white flour
>   1/2 ounce (2 envelopes) dry yeast
>   2 1/2 cups lukewarm water (110–115° F)
>   1 tablespoon salt
>   Flour for dusting

To Make the Starter
In a non-metallic mixing bowl, add yeast, sugar, and mix with lukewarm water. Gradually add flour, mixing to form a soft dough.

Remove dough from mixing bowl, place onto a flat surface, and knead for 7-10 minutes. Wipe bowl clean and return dough into the bowl.

Cover dough with oiled plastic wrap and a kitchen towel, and let sit at room temperature for 3-4 hours, or until well-risen and starting to collapse.

In a large non-metallic mixing bowl, mix the 2 yeast envelopes into 1 cup of the lukewarm water until yeast is dissolved; stir in the remaining water. This will be a very watery batter.

Add previously-made starter; mix together until starter is dissolved. Gradually add flour and salt; mix to form a soft dough.

On a lightly floured surface, knead the dough for 10 minutes until it becomes smooth and elastic. Place the dough into an oiled bowl, cover with oiled plastic wrap and a kitchen towel, and let rise in a warm place for about 2 hours, or until it doubles in volume.

Punch dough down and on a lightly-floured surface, knead the dough for a couple of minutes. Cut dough in half and shape each half into round balls; flatten slightly.

Place loaves onto oiled baking sheets, cover with oiled plastic wrap and a kitchen towel, and let rise in a warm place for 1 hour, or until it almost doubles in volume.

Preheat oven to 450° F. Place a bowl of boiling water on the bottom oven rack to create moisture.

Dust the top of the dough with flour, and with a very sharp knife, score the top of each loaf about 3/4 inches deep in a crisscross pattern.

Bake bread for 30 minutes, or until tops are golden. Remove from oven and set aside to cool.

Preparation time: 30 min.   Baking time: 30 min.   Yield: 2 loaves

New Year's Pita Bread
(Vassilopita)

This bread is traditional for New Year's Eve. Everyone bakes this pita bread and incorporates a gold coin into it. During dinner, the pita bread is cut into individual pieces. The luckiest person is the one who finds the coin.

    2 pounds all-purpose flour
    1 ounce (4 envelopes) active dry yeast
    1/2 cup lukewarm water (110–115° F)
    1 teaspoon ground mahlepi
    1 teaspoon ground kakoule (optional)
    1 teaspoon salt
    4 large eggs, room temperature
    1 1/2 cup (12 ounces) sugar
    1 egg white (keep yolk for decorating)
    Decoration
    1 tablespoon milk
    1 tablespoon sesame seeds
    2 tablespoons almonds, blanched, sliced

Optional Decoration
    2 tablespoons flour
    4–5 tablespoons ouzo (or vodka)
    1/2 teaspoon sugar
    1 cup lukewarm milk
    8 ounces butter, melted
    1–3 coins (optional)

In a medium, non-metallic bowl, dissolve yeast in 1/2 cup lukewarm water. Let rest for 30 minutes until it foams and doubles in size.

Mix flour, mahlepi, kakoule, and salt; sift together 3 times. In a large bowl, combine eggs, sugar, and egg white; mix until smooth. Add yeast; mix well. Gradually add flour mixture; mix well. Dough will be very hard. Slowly add remaining milk, and last, add butter; mix. Don't overwork dough.

Cover with oiled plastic wrap and a kitchen towel; set aside at room temperature to double in volume. This takes approximately 3–4 hours.

Punch dough down; divide into 2 or 3 pieces. On a lightly-floured surface, fold dough 2-3 times; place into 10 x 2-inch round, buttered, oven pans. If you are making this for a New Year's Eve celebration, it is customary to place a coin into each pita for good luck. Wrap coins in aluminum foil and work into dough.

Cover with oiled plastic wrap and a kitchen towel; let rise for approximately 1 hour, or until dough doubles in volume.

Preheat oven to 350° F. Beat egg yolk with 1 tablespoon milk; gently brush over dough. Sprinkle on sesame seeds, and decorate with almonds.

To write a message on the bread, mix 2 tablespoons flour, ouzo, and sugar (add water, if needed), to make a thick, creamy paste, like mayonnaise. Use a decorating bag and a pipe to write the message on top of the pita. Do not brush decoration with egg. You want the message to stay white in contrast to the brown surface of the pita.

Bake on middle oven rack for 40–45 minutes, or until golden brown. Set aside to cool.

Tip: To dissolve yeast, water should not be over 115° F; hotter temperatures will "kill" the yeast. Active yeast bubbles and foams within minutes. Replace yeast, if necessary. Let dough rise at a room temperature of about 75° F, or otherwise it will take longer. If dough is sticky, add flour to your hands. Do not add flour to the dough.

Preparation time: 30 min.   Baking time: 40 min.   Yield: 2–3 Vassilopitas

Easter Bread
(*Tsourekia*)

This is a very aromatic bread. You can make it any time and it is great for breakfast with some butter and jelly or just dip it in your coffee. Leftover bread makes wonderful French toast.

2 pounds all-purpose flour
1 1/2 ounces (6 envelopes) active dry yeast
1 1/2 cups lukewarm milk (110–115)° F
1/4-teaspoon ground Mastich
1 teaspoon ground mahlepi
1 teaspoon salt
5 large eggs, room temperature
1 egg white (keep yolk)
1 1/2 cups (12 ounces) sugar
1/2 teaspoon orange extract
8 ounces butter, melted

Decoration
1 tablespoon milk
1 egg yolk
4 tablespoons almonds blanched, halved
4 eggs, hard-boiled and colored red (optional)

In a small bowl, dissolve yeast in milk. Milk must be warm enough to activate the yeast. Let stand for 30 minutes until it foams and doubles in size.

In large mixing bowl, combine flour, mastich, mahlepi, and salt; sieve 3 times.

Beat eggs, egg white, sugar, and orange extract until smooth. Add yeast and mix; gradually add flour, mixing to incorporate the ingredients.

Add butter. Mix dough until all butter has been incorporated. Do not overwork dough. Butter will be absorbed by the flour as it rises.

Cover with lightly-buttered plastic wrap and a kitchen towel; set at room temperature (75° F) to double in volume (approximately 3-4 hours). Grease two 15 x 12 x 1-inch baking sheets and set aside.

When dough has doubled in volume, punch dough to collapse and place on a lightly-floured, flat surface. Separate dough into 12 equal pieces.

Shape pieces into 9-inch long strips (like 1-inch thick sausages). All pieces should be the same size. Pinch 3 strips together at one end then braid them neatly. Pinch the ends together and tuck under the braid.

Place 2 braids on the prepared baking sheet, spacing them well apart. Cover with buttered plastic wrap and a kitchen towel; let rise at room temperature for 60 minutes or until they double in volume.

Preheat oven to 350° F. Combine egg yolk and milk; brush over braids. Arrange some almonds on top. Put colored egg, if using, in the middle of each braid; push egg in so that dough covers half the egg.

Bake on middle oven rack for approximately 30 minutes, or until golden.

Remove from oven and transfer to a wire rack to cool.

Tip: To dissolve yeast, milk should be warm, but not over 115° F; hotter temperatures will "kill" the yeast. Active yeast bubbles and foams within minutes. Replace yeast, if necessary. If dough is sticky, add flour to your hands. Do not add flour to the dough.

Preparation time: 30 min.   Baking time: 30 min.   Yield: 4 Tsourekia

Red-Colored Eggs
(Pashalina Kokkina Avga)

This recipe is the traditional way of coloring eggs in Mytilini for Easter. You will see many different colors but the most popular color is red. The eggs are dyed on Holy Thursday and they are used to decorate the Easter table and Easter Bread (Vassilopita).

 1 dozen eggs
 1 package powdered red dye (use color of your choice).
 1/2 cup white vinegar

In a medium-size saucepan place the eggs and pour enough water to cover the eggs. Dilute dye in vinegar and add to the water.

Slowly bring to a boil and simmer for about 8 minutes, stirring occasionally. Remove eggs and cool.

Tip: To give a shiny look to the eggs, when slightly cooled, rub each egg with a lightly-oiled cloth. Eggs should be at room temperature before boiling to prevent them from cracking.

Preparation time: 10 min.   Cooking time: 8-10 min.   Yield: 1 dozen

GLOSSARY

Avgolemono—A sauce made with egg and lemon juice thickened with cornstarch and combined with a broth of your choice. This sauce is a favorite in Greek cooking.

Bakala—Salted codfish, primarily sold in the winter months. Bakala needs to be soaked in water for 24 hours before it is to be cooked, as it is very salty.

Baklava—This is the most famous and best-known Greek dessert. It is made in many different ways, but the basic ingredients are: very fine filo pastry, walnuts, almonds or pistachio nuts, and syrup.

Fava Beans—These beans can be found fresh in the pod and green like a flat string bean, or dry (just the seed pod). They grow in mid to late winter. They look like flat string beans, but they are not string beans. They are delicious cooked fresh with artichokes, or dry with olive oil.

Feta Cheese—Feta is the best-known Greek cheese. It is a semi-hard, tangy, white cheese. It has salty flavor, as it is cured for weeks or months in salt brine. It is made from goat or sheep milk. Commercially, you will find Feta that is made out of pasteurized cow's milk.

Filo Dough or Filo Pastry—There are three or four different thicknesses of filo dough available on the market. I prefer to use #4 in my recipes. It is sold in 1-pound packages. Each package contains about 25 to 27 count of about 17 x 12-inch paper-thin sheets, wrapped in a sealed plastic foil. This dough dries up very quickly, so you will need to keep it covered with a damp (not wet) kitchen towel when you are working with it. It can be found in the frozen section of most supermarkets. Defrost filo dough for at least 12 hours before using. If you have the time, defrost for 24 hours.

Grapevine Leaves—These are sold in supermarkets. They are preserved in brine and neatly packaged in jars. You can use fresh grapevine leaves if you or your neighbor has a vine tree. Fresh ones are at their best before they become fuzzy and hard.

Graviera—A cheese with a sweet, fruity flavor that is made from sheep's milk. It is an excellent cheese to serve as an appetizer or after dinner with fresh or dry fruit.

Htenia—A type of small shellfish that resembles scallops.

Kadaifi Pastry—This pastry is available in Greek or Middle Eastern stores or via the Internet. It is shredded pastry and looks like angel hair (it is not pasta), but it is soft. It dries up very quickly; therefore, you will need to keep it covered with a damp (not wet) kitchen towel. Defrost for 4 to 6 hours before using.

Kasseri—A semi-hard cheese with a light, creamy, golden color, and a mild flavor. It is excellent for saganaki (fried cheese).

Kefalotyri—A hard, pale yellow cheese with a tangy, salty flavor. It resembles pecorino and Romano cheese.

Kefalograviera—A semi-hard cheese made from sheep's milk. It is not as salty as the Kefalotyri cheese. Excellent as a table cheese or to make saganaki (fried cheese).

Mastich—The resin of a shrub grown on the island of Chios. The resin hardens to form crystals. Grind the crystals into a powder before you use it in the various recipes.

Mahlepi—Used in powder form, mahlepi is produced from the inner kernels of the fruit pits from a native Middle Eastern cherry tree. The kernels are dried and ground to a fine powder. This powder is used in numerous recipes.

Meze—The Greek word for hors d'oeuvres. Greeks enjoy their meze at lunch or dinner. Sometimes meze becomes the entire meal.

Myzithra—A white cheese that is similar to ricotta cheese. It can be eaten as fresh or dry, grated like Parmesan cheese.

Nerantzi—Bitter orange. It looks like an orange, but it has a sour and bitter flavor. Sometimes it is used as a substitute for lemon juice. Young and small bitter oranges the size of a walnut are used to make the well-known preserve called nerantzaki glyko.

Opa—This is what your waiter will announce when a saganaki (fried cheese) dish is ignited. It is an exclamation of appreciation for having fun and good times.

Orzo—A pasta used in soups or main dishes. The most famous dish with orzo is giouvetsi.

Ouzo—This is the most popular drink on the island of Mytilini. It has been produced there for a very long time. Made with anise, ouzo is a clear alcoholic beverage that is mixed with water before consumption. When water is added, ouzo obtains a milky color. You should never add ice directly into ouzo. The best way to enjoy ouzo is to always add cold water, to indulge in a good meze, and to be in the company of family and good friends.

Tarama—Carp roe eggs, salted and aged. The eggs are small and a light pink in color. It is sold in small jars.

INDEX

Favorite Recipes & Notes